Guide to Advanced Statistical Analysis in R

currently in the Statistics without Mathematics series

Guide to Advanced Statistical Analysis in R

Advanced data analysis – without tears

April Liu

Statistics without Mathematics series

General Editor
Cole Davis

Vor Press

Contents

Foreword

by the General Editor

Introduction

Statistics is a seemingly very mysterious yet necessary subject at graduate level. Many graduate students are required to use statistics to carry out their research, be their study in science, medicine, engineering, business, or social sciences. Most universities provide basic courses in statistics for students at undergraduate and graduate level, including research design and the analysis of data. However, most courses are not long enough to cover statistics beyond a scattering of basic tests, and more advanced statistical methods are usually not explained in such a way as to be understood by novice statistics students, especially those without a mathematical leaning.

How does this book teach statistics?

In common with other books in the 'Statistics without Mathematics' series, each test is accompanied by a worked example. In particular, April Liu gives a running explanation of how the R functions are used, so that relatively new users of R should be able to dip into any chapter and reuse the code therein to examine their own datasets. She also recommends reading materials should the reader wish to study a test in greater depth. It should be emphasized that this book keeps it light, superficial even, in order for the test user to get started on data analysis with advanced statistical methods without becoming bogged down in theory and equations. April explains any complexities of the test in simple language which a non-statistician can easily follow.

The contents of the book

It could be argued that this book should be called *Beyond Regression*, in that many of the tests included here are devoted to doing things which multiple regression cannot, or building on top of its magnificent edifice.

Chapter 1 – Structural Equation Modeling

includes something distinctly missing from standard multiple regression, that of latent (non-measurable) variables and manifest variables. Within this particular body of statistical knowledge, Dr Liu builds up a cogent system for data analysis by dividing the chapter into sub-techniques (each of which may be used on its own). **Path analysis** is a study of causal directions. **Confirmatory Factor Analysis** allows refinement of established causal models, seeing how well latent variables may be inferred; as well as the basic usage of CFA, Liu provides an example of Multiple Groups CFA, comparing structural equation models between different groups in a dataset. **Basic SEM** builds on path analysis and CFA by testing for a significant relationship between latent variables. **Latent growth models**, although under the SEM umbrella, analyze repeated measures (longitudinal data).

Chapter 2 – Time Series Analysis

deals with linear, univariate time series. Unlike linear regression, time series models do not separate outcome and independent variables, but view the same variable at different points in time. In particular, Liu deals with the issue of *non-stationary* versus *stationary data*, including transforming the former into the latter, and plotting the *rolling mean*. She demonstrates the use of **ARIMA modeling**, **Auto ARIMA**, and **Seasonal ARIMA** (SARIMA).

Chapter 3 – Survival Analysis

is also devoted to time, but in a very specific way. It analyses time-to-event data, how much time it takes to reach a certain event, whether negative (such as death) or positive. The first type of survival analysis included here is the **life table**, also known as the actuarial table. Probably rather simpler to use and with very clear results is the **Kaplan-Meier** method (also known as Product limit method), which creates a curve, allowing ease of interpretation and readily providing fresh questions for the researcher. The **Cox model**, or Cox proportional hazard model if we feel wordy, is a semi-parametric test allowing the examination of multiple factors influencing survival time. Two parametric survival methods are also introduced: the **Weibull** and **exponential** distributions.

Chapter 4 Longitudinal Analysis

which could also be titled repeated measures or panel analysis, is also time-related. Its purpose is distinct from time series, in that longitudinal analysis is designed to examine differences between variables collected at different times, and the influence of other variables in the dataset. Starting with **repeated measures ANOVA**, the chapter goes on to study the **linear mixed effects model** – which distinguishes between fixed and random effects – and then **generalized estimating equations** (GEE), which model population averages.

Chapter 5 – Multivariate Analysis

may also be called multi-outcome analysis; this alternative title, while less commonly used, is perhaps more helpful in understanding this branch of statistics. The first method discussed, **discriminant analysis**, allows the categorization of cases (or individuals) in a dataset into different groups based on the differences between their characteristics. It has similarities to multiple regression in that it calculates weights (the numerical impact) in order to make its predictions. Three methods of discriminant analysis are shown, linear, quadratic, and regularized, although the reader is made aware of other methods and relevant functions in R. Another analysis is **canonical correlational analysis**, which identifies and measures the associations among two sets of variables in a dataset. In addition to predictions, we may also find out about the number of significant dimensions within the dataset (thus being similar to multiple regression but with an extra step). The concluding method in this chapter is **multidimensional scaling** (MDS), which allows visualisation of the distances (level of similarity or difference) between subjects; metric and non-metric MDS are shown.

Chapter 6 – Miscellaneous Methods

is of course a mixed bag. Each topic is important, but none are individually broad enough to merit a separate chapter in the survey of methods that comprises this book. The first section is devoted to **Generalized Linear Models** (GLM) and **Poisson regression**. GLM allows flexibility in studying datasets which fail to meet the assumptions for data of linear models. The most well-known of these are logistic regression, covered in this volume's sister book (see footnote at the end of the chapter), and discussed in this volume, Poisson regression, where the data comprises positive whole numbers (e.g. car crashes per year, new employees per month, cases of covid per week). The next section of the chapter is about **hierarchical modeling**, which particularly deals with cases where the independence assumption for data has not been met; also known as nested or multilevel modeling, it is particularly useful for studies which cover hierarchical groups, such as classes within schools and schools within regions, or wards within hospitals and hospitals within health authorities. The next section deals with **power analysis**; in particular, we can calculate suitable sample sizes, in the hope that we can avoid statistical errors; a set of functions in R are recommended, which can be applied to a wide range of statistical tests. While the subject of **reliability** is situated last in the book, this does not make it unimportant: without reliable measures, tools of prediction can be worse than inaccurate, they can actively mislead. April shows methods for examining the reliability of data in three separate ways: test-retest reliability, internal reliability (also known as internal consistency), and inter-rater reliability.

What's missing from the book?

Clearly, some advanced tests will not be available in one volume, especially as this is a book dedicated in particular to non-mathematicians and non-statisticians. However, some advanced tests which do not always feature in introductory books may be found in this book's sister volume (see the footnote at the end of this chapter), which takes the reader from beginner to intermediate level. The other book includes categorical tests, such as the binomial test, multinomial test (also known as Chi squared Goodness of Fit), and log-linear analysis. It also features logistic regression, MANOVA, principal components analysis, exploratory factor analysis, cluster analysis and an introduction to Bayesian statistics.

About the author

April Liu holds a PhD in Biostatistics from the University of Saskatchewan in Saskatoon, Saskatchewan, Canada and has spent the last 3 years as a statistician in the healthcare industry. She currently lives in Ottawa, Ontario, Canada.

Text formats

Ordinary text in the book looks like this:
Yes, just like this.

The following is input code which you type into the R command line. The hash symbol (#) precedes inline comments, which are not processed except as output on the monitor:

```
newFile = read.csv("Social.csv") # read file,
    # allocated to the new object, newFile
head(newFile) # Let's look at the first six lines
```

and below are the results, typical data output from R:

```
  id   extro    open    agree    social class school
1  1 63.69356 43.43306 38.02668  75.05811     d     IV
2  2 69.48244 46.86979 31.48957  98.12560     a     VI
3  3 79.74006 32.27013 40.20866 116.33897     d     VI
4  4 62.96674 44.40790 30.50866  90.46888     c     IV
5  5 64.24582 36.86337 37.43949  98.51873     d     IV
6  6 50.97107 46.25627 38.83196  75.21992     d      I
```

Cole Davis

General Editor
Statistics Without Mathematics series

Norwich, November 2022

Davis C (2022) *Statistical Testing with R* (second edition). Norwich: Vor Press.

Chapter 1 – Structural Equation Modeling

Sections include

Path Analysis

Confirmatory Factor Analysis
Basic CFA
Multiple Groups CFA

Basic SEM

Latent Growth Models

Introduction

Imagine that you are a researcher who is looking to see if there is a relationship between the level of religiousness of people and their level of emotional health; or imagine that you are an educator who wants to understand whether your teaching style and schemes of work can improve students' success and whether the effects will differ among students of different social economic status? If you had neither training nor exposure to statistics before you picked up this book, at first glance you might think that these issues are very different and require different statistical methods to resolve them. However, if you have some statistical training (say, if you learned about regression before reaching for this book), you might say "April, don't patronize me with these simple problems! For the religion and emotional health problem, I can use emotional health as the outcome variable and religiousness as the independent variable. For the educator problem, I can use standardized test grades as the outcome variable, and teaching style, scheme of work, and socioeconomic status of students as independent variables."

The student trained in regression missed an important detail. A person's religiousness, or emotional health, or academic success, or need (or lack of it) for power tools, are not variables that are measurable. They are what statisticians call **latent** variables (aka non-measurable or non-observable variables) because they can only be inferred from variables that are measurable. A measurable variable (aka **manifest** or observable variable) is a variable which can be observed and directly measured. For example, we may not be able to directly measure the latent variable 'Success' because there are many ways to define success (monetary success, social popularity, emotional happiness, etc), but we can infer it with measurable variables such as salary, number of close friends, and self-reported level of happiness. Structural Equation Modeling (SEM) is a branch of statistics that is used to understand the relationship between measurable and non-measurable variables as well as non-measurable variables with other non-measurable variables. There are four main methods that fall under the SEM category: Path Analysis; Confirmatory Factor Analysis (CFA); Basic SEM; and Latent Growth Modeling (LGM).

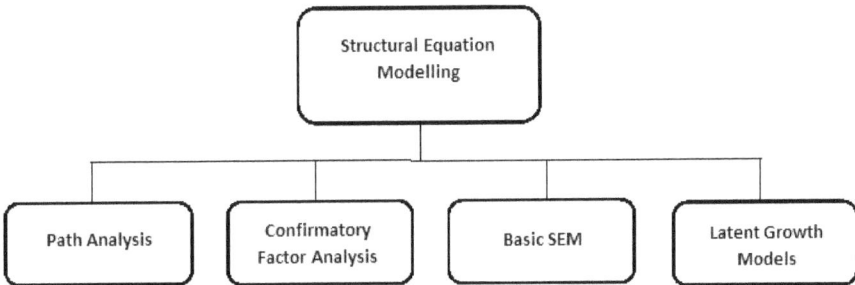

```
                    ┌─────────────────────┐
                    │ Structural Equation │
                    │     Modelling       │
                    └─────────────────────┘
       ┌──────────────┬──────────┴────────┬──────────────┐
┌──────────────┐ ┌──────────────┐ ┌──────────────┐ ┌──────────────┐
│              │ │ Confirmatory │ │              │ │ Latent Growth│
│ Path Analysis│ │Factor Analysis│ │  Basic SEM   │ │    Models    │
└──────────────┘ └──────────────┘ └──────────────┘ └──────────────┘
```

Path Analysis and CFA are two analyses which should be performed before performing Basic SEM and LGM to ensure the accuracy of the results from the last two methods. Therefore, if SEM is a completely new topic for you, you may find it useful to read the sections on Path Analysis and CFA and perform these methods before attempting more complex methods.

Path Analysis

Before beginning any SEM, it is useful to learn first about the causal connections of the variables in the dataset of interest; path analysis is a great method for this. In order to perform path analysis, two important requirements must be fulfilled by the variables in the dataset. First, each variable must have an influence on another one *in a single direction* of causation. For example, the length of a person's education may influence salary, but not the other way around (Schumacker and Lomax, 2010). Second, an event of a variable that is believed to have a causal relationship with another variable must precede the event of the other variable. To reuse the example of length of education and salary, a person's salary can only be influenced by his/her education obtained before the salary, not afterwards (Schumacker and Lomax, 2010). Additionally, in order to perform path analysis accurately, the dataset must meet the multivariate normal assumption as well as having a sample size of 5 - 10 individuals for each observed variable in the analysis (Klem, 1995).

As is usual with statistical methods, the best way to explain path analysis is through an example. The World Happiness Report 2015 dataset was collected from 156 countries around the world. It contains the happiness score of each country as well as scores for variables such as the economy, existence of family and friends, freedom, and trust in the government (Helliwell *et al*, 2015).

The R package *lavaan* is a great package for implementing all of the methods under the SEM umbrella. The first thing we should do is to install the *lavaan* package using

Path Analysis

the install.packages() function, and after we finish installing *lavaan*, we can use the library() function to load all of the *lavaan* package's functions (Rosseel, 2012) within the R environment.

```
# World Happiness 2015 example
install.packages("lavaan")
library(lavaan)
```

It is then time to read the dataset into the R environment by using the read.csv() function (as the dataset is stored in .csv format). The dataset is assigned to the object named WorldHappinessData2015:

```
WorldHappinessData2015 <-
    read.csv("WorldHappinessData2015.csv")
```

For path analysis, our first step is to hypothesize the relationships of the variables within the dataset before we use the method to test for the statistical significance of these relationships. Let's say that you hypothesized that the happiness scores of countries around the world depend on their economy, the companionship of family and friends, life expectancy, freedom, and trust in the government, while trust in the government depends on the freedom and economy of the countries. You can write these hypothesized relationships into structural equations (or structural models) using the names of the variables in the dataset within a string. You assign the string to an object with the name of your choice; here, we choose the name *model*.

```
model <- '
    HappinessScore ~ Economy + Family
    + LifeExpectancy + Freedom
    + GovernmentTrust

    GovernmentTrust ~ Freedom + Economy
    '
```

From the equations specified in the string, you can see that there are tildes (~) between the dependent and independent variables, which represent causal relationships between the variables. HappinessScore and GovernmentTrust are the dependent (or outcome) variables. The other variables are the independent variables.

After the model has been specified, the next step is to test and see if the theorized model correctly matches the variance-covariance matrix of the dataset. This is where the actual path analysis is performed! To do this, we use the cfa() function of the *lavaan* package and we enter the name of the structural equation (*model*) as the first argument and the name of the dataset as the second argument; the entire fitting of the model is assigned the name *fit*.

```
fit <- cfa(model, data = WorldHappinessData2015)
```

To see the results of the path analysis, apply the summary() function to fit:

```
summary(fit, fit.measures = TRUE,
    standardized=TRUE, rsquare=TRUE)
```

The output shown comprises a selection from a quite verbose output:

```
lavaan 0.6-9 ended normally after 35 iterations

  Estimator                                    ML
  Optimization method                      NLMINB
  Number of model parameters                    9

  Number of observations                      158

Model Test User Model:

  Test statistic                            3.317
  Degrees of freedom                            2
  P-value (Chi-square)                      0.190
```

The chi-square test is used to determine how well the theorized structural equations fit the variance-covariance matrix of the dataset. The null hypothesis is that 'theorized structural equations fit the variance-covariance matrix of the dataset'. In this example, from the Model Test User Model section of the output, you can see that the p value for Chi-Square is 0.190, which means we cannot reject the null hypothesis at the 95% confidence level. The null hypothesis is that the theorized structural equations are accurate: happiness within countries is dependent upon the economy, family-friendliness, life expectancy, freedom, and trust of government; trust of government is dependent on freedom and the economy of the countries, according to the World Happiness Dataset 2015 and its variance-covariance matrix.

```
Regressions:
                    Estimate  Std.Err  z-value  P(>|z|)   Std.lv  Std.all
  HappinessScore ~
    Economy           0.805    0.206    3.903    0.000    0.805    0.283
    Family            1.416    0.216    6.547    0.000    1.416    0.336
    LifeExpectancy    1.034    0.304    3.398    0.001    1.034    0.222
    Freedom           1.443    0.355    4.062    0.000    1.443    0.189
    GovernmentTrst    0.854    0.418    2.042    0.041    0.854    0.089
  GovernmentTrust ~
    Freedom           0.350    0.059    5.977    0.000    0.350    0.440
    Economy           0.043    0.022    1.971    0.049    0.043    0.145
```

From the Regressions section, the estimate values are the estimates of the parameters of the structural equations, representing the amount of causal effect the variables have on citizens' happiness and their trust in the governments. From the p values of these estimates, you will see that in this example, all the variables have significant effects on the citizens' happiness and their trust in their governments.

An easier way of seeing the relationships between the variables of theorized structural equations after the fit has been created is through a path diagram. The *lavaanPlot* R package contains a *lavaanPlot()* function which can construct the path diagram of the fitted structural equations (Lishinski, 2018). We use the install.packages() and library() functions as before with the lavaanPlot package. The lavaanPlot() function generates the path diagram by using the information from the fitted structural equations, using 'fit' as created earlier. Arguments may be customized to some extent.

```
install.packages("lavaanPlot")
library(lavaanPlot)
lavaanPlot(model = fit,
    node_options = list(shape = "box",
```

```
fontname = "Helvetica"),
edge_options = list(color = "grey"),
coefs = TRUE, sig=0.05,
stars = c("regress"))
```

The first argument, *model*, uses the name we assigned to the fitted structural equations, *fit*. A path diagram is composed of the variables involved (nodes) and their causal relationships (arrows or 'edges'). The options may be customized to some extent. You can change arrows to red, for example, and turn off coefficients by coefs = FALSE. The *sig* argument restricts the appearance of causation coefficients to a certain significance level, in this example 0.05. The *stars* argument places a certain number of stars beside each causation coefficient, the number of stars representing the level of significance.

This image should appear in your browser:

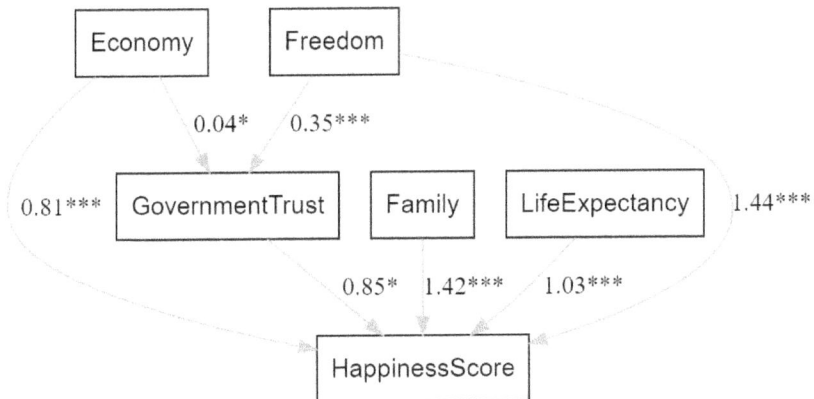

Dependent variables (here, HappinessScore and GovernmentTrust) are also called endogenous (meaning 'internal')

variables because they are caused by at least one other variable within the model; they have arrows pointed towards them by the other variables. The independent variables in both equations (Economy, Family, LifeExpectancy, and Freedom) can also be called exogenous ('external') variables because they are not caused by any variables in the equation, but they do exert a causal influence on at least one variable in the equation.

GovernmentTrust's causation coefficient on HappinessScore is 0.85 with a p value of 0.041 (as read from the Regressions statistics). Although 0.041 is below 0.05, it is very close to it, therefore only 1 star is placed beside 0.85. For a more significant causation such as Freedom on GovernmentTrust, the causation coefficient of 0.35 had a p value of 0.000 (which means many decimal places are zeros before we have a non-zero decimal place), therefore 3 stars are placed beside 0.35.

When you have structural equations containing variables that do not have significant causal effects on your dependent variables, the chi-square (the test statistic) value can be larger, sometimes forcing you to reject the null hypothesis. Part of the art of path analysis is to keep variables with significant causal effects on your dependent variables of interest and to remove non-significant variables in order to find the best structural equations for the dataset. If you can master this skill, even if the structural equations you theorized in the beginning are wrong, you can still find structural equations which are close to your original theorized equations. You can start your journey of mastering path analysis by testing out your own theorized structural equations on the data set we have used and on other sample datasets.

Confirmatory Factor Analysis

When you hear the term Confirmatory Factor Analysis, it may remind you of another similar factor analysis method called Exploratory Factor Analysis (EFA). For EFA, you begin the analysis without knowing how many latent variables may be inferred from the observed variables in the dataset. Unlike EFA, CFA assumes that you already have a solid idea of what and how many latent variables may be inferred from the observed variables in the dataset (Davis, 2022). A factor model is then imposed upon the data, testing a theory and evaluating its goodness of fit to the data (Bryant and Yarnold, 1995).

Like path analysis, CFA also assumes multivariate normality and the sample size rule of 5 – 10 individuals for each observed variable (Bryant and Yarnold, 1995). Also like path analysis, it is wise for this analysis to be performed on the dataset before attempting SEM because it is a method designed to *confirm* whether the observed variables are good for inferring the latent variables to be used in SEM (Schumacker and Lomax, 2010).

When you are analysing a dataset using SEM, the latent variables are never in the dataset; they are usually theorized by analysts like yourself before the analysis. There are a few different types of CFA which cater for different types of dataset and analysis needs. For this book, we will go through Basic CFA and Multiple Groups CFA.

Basic CFA

This section will give you a very basic understanding of how CFA works. You will guess from its name that the method is often used for relatively simple datasets that were collected from a single population where differences between the individuals cannot significantly influence the result of your intended analyses. This example uses World Happiness Data from 2019. The dataset, like that of 2015, contains happiness data collected from 156 countries. But unlike the 2015 dataset, the 2019 dataset contains the level of social support provided by the country instead of the family-friendliness of the country, and people's perception of the country's levels of corruption in place of the trust the people place in their governments (Helliwell *et al*, 2019). In this example, we are going to look at whether or not the National Level of Happiness may be inferred by GDP, social support, and life expectancy, and whether or not the Personal Level of Happiness may be inferred from freedom, generosity, and perceptions of corruption.

```
# World Happiness 2019 example
library(lavaan)
WorldHappinessData2019 <-
    read.csv("WorldHappinessData2019.csv")
```

As with path analysis, we load the functions in the lavaan package for this analysis and use the read.csv() function to read the WorldHappinessData2019 data into the R environment. After that, we need to cut the dataset to include only columns 4 – 9 of the dataset because only these variables will be used in the example analysis.

```
WorldHappinessDataCut <-
    WorldHappinessData2019[,4:9]
```

The right-hand side of the comma in the square brackets selects the columns; in this case, 4:9 selects columns 4 to 9. The left-hand side of the comma can be used to select rows in the same manner. WorldHappinessDataCut is now a data frame with the selected columns. We next convert the data frame into the variance-covariance matrix format, necessary for running CFA, using the cov() function. The variance-covariance matrix is assigned to the object named WorldHappinessCov:

```
WorldHappinessCov <- cov(WorldHappinessDataCut)
```

Next, we specify the structural equations! For CFA, between the latent variables being inferred and the observed variables inferring them is the sign '=~' which stands for 'infers'. Beneath the two specified equations for inferring latent variables is an equation which represents the possible correlation between the two latent variables; to show the correlation, there is a double tilde (~~) between them for this purpose. We assign these three structural equations to the name *model1*.

```
model1 <- '
  national =~ GDP.per.capita
    + Social.support + Healthy.life.expectancy

    personal =~ Freedom.to.make.life.choices +
      Generosity + Perceptions.of.corruption

    personal ~~ national
    '
```

After we finish specifying the structural equations, as in path analysis, we use the cfa() function to test whether or not the relationships specified by the model fit the data (in this case, the variance-covariance matrix of the data).

```
cfa.model <- cfa(model1,
    sample.cov= WorldHappinessCov,
    sample.nobs = 156, std.lv = TRUE)
```

In the first argument of the model, we put the name of the structural equation 'model1', the second argument, sample.cov, holds our variance-covariance matrix, and the third argument sample.nobs is the number of individuals in the dataset. Because CFA is a factor analysis, we add the argument std.lv = TRUE to the cfa() function to make sure that the factor loading (the coefficients of observed variables in the structural equations of a CFA) of the first observed variables in each structural equation are not standardized to 1. The entire test is assigned the name cfa.model and we can look at the results by using the summary() function on cfa.model:

```
summary(cfa.model)
```

```
Model Test User Model:

  Test statistic                        33.498
  Degrees of freedom                         8
  P-value (Chi-square)                   0.000


Latent Variables:
                   Estimate  Std.Err  z-value  P(>|z|)
  national =~
    GDP.per.capita    0.369    0.025   14.804    0.000
    Social.support    0.242    0.020   12.009    0.000
    Hlthy.lf.xpctn    0.216    0.016   13.948    0.000
  personal =~
    Frdm.t.mk.lf.c    0.114    0.015    7.664    0.000
    Generosity        0.031    0.009    3.514    0.000
    Prcptns.f.crrp    0.055    0.009    6.115    0.000
```

To see the path diagram of the CFA model, we can use the lavaanPlot() function from the lavaanPlot package, as in Path Analysis:

```
library(lavaanPlot)
lavaanPlot(model = cfa.model,
    node_options = list(shape = "box",
    fontname = "Helvetica"),
    edge_options = list(color = "grey"),
    coefs = TRUE, covs=TRUE,
    stars = c("latent", "covs"))
```

The arguments in the lavaanPlot() function are similar to those for Path Analysis. However, for CFA the coefs=TRUE argument means that the estimates of the factor loadings in the latent variables section of the summary output are placed beside the arrows. Also unlike in the Path Analysis example, another argument needs to be included, covs=TRUE, to include the covariance between the latent variables. The argument 'stars' is also different for CFA than for Path Analysis: instead of including the string "regress" for the regression section, we use a vector with the "latent" and "covs" strings, representing the factor loadings and covariances from the Latent Variables and Covariances sections of the summary output.

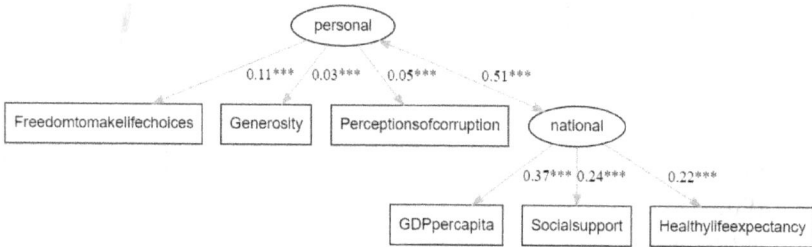

Between 'personal' and 'national', we see a double arrow representing covariance. The covariance between the latent variables in this example has a value of 0.51.

From the results, we can see that although all of the observed variables have significant factor loadings in the structural equations, the chi-square test p value is smaller than 0.05, which means the structural equations do not fit the dataset. This usually happens in situations where there are additional correlations between the variables used in the structural equations that need to be accounted for. To find these correlations, we can use the *modindices*() function on the cfa.model and include the necessary correlations to find the structural equations that suit the dataset.

```
options(digits=3)
# To avoid verbose numerical output
options(width=110)
# To present broader output
modindices(cfa.model)
```

	lhs	op	rhs	mi	epc	sepc.lv	sepc.all	sepc.nox
16	national	=~	Freedom.to.make.life.choices	7.912	0.091	0.091	0.635	0.635
17	national	=~	Generosity	14.464	-0.038	-0.038	-0.401	-0.401
18	national	=~	Perceptions.of.corruption	0.201	0.006	0.006	0.067	0.067
19	personal	=~	GDP.per.capita	1.564	-0.032	-0.032	-0.081	-0.081
20	personal	=~	Social.support	1.761	0.030	0.030	0.100	0.100
21	personal	=~	Healthy.life.expectancy	0.071	0.004	0.004	0.018	0.018
22	GDP.per.capita	~~	Social.support	0.071	0.003	0.003	0.100	0.100
23	GDP.per.capita	~~	Healthy.life.expectancy	1.761	0.014	0.014	0.918	0.918
24	GDP.per.capita	~~	Freedom.to.make.life.choices	2.270	-0.003	-0.003	-0.245	-0.245
25	GDP.per.capita	~~	Generosity	3.198	-0.003	-0.003	-0.201	-0.201
26	GDP.per.capita	~~	Perceptions.of.corruption	2.044	0.002	0.002	0.175	0.175
27	Social.support	~~	Healthy.life.expectancy	1.564	-0.007	-0.007	-0.355	-0.355
28	Social.support	~~	Freedom.to.make.life.choices	11.518	0.006	0.006	0.412	0.412
29	Social.support	~~	Generosity	0.442	-0.001	-0.001	-0.058	-0.058
30	Social.support	~~	Perceptions.of.corruption	7.574	-0.003	-0.003	-0.257	-0.257
31	Healthy.life.expectancy	~~	Freedom.to.make.life.choices	0.127	0.000	0.000	-0.049	-0.049
32	Healthy.life.expectancy	~~	Generosity	0.051	0.000	0.000	0.022	0.022
33	Healthy.life.expectancy	~~	Perceptions.of.corruption	0.560	0.001	0.001	0.078	0.078
34	Freedom.to.make.life.choices	~~	Generosity	0.201	0.001	0.001	0.078	0.078
35	Freedom.to.make.life.choices	~~	Perceptions.of.corruption	14.464	-0.011	-0.011	-1.720	-1.720
36	Generosity	~~	Perceptions.of.corruption	7.912	0.002	0.002	0.283	0.283

From the output of the modification indices, we can look through the correlation relations where there are two tildes (~~) and find the relationship with the highest mi value which also has a positive epc. In this case, the correlation relationship with the highest mi value is between social support and freedom. We then add this correlation relationship to the specified structural equations and use the cfa() function see if the equations now fit the variance-covariance matrix of the dataset.

Sometimes, the process of modifying the structural equations and checking the modification indices must be done multiple times before the p value of the chi-square test can be above 0.05. In the example of WorldHappiness-

Data2019, the additional correlations added to the structural equations are social support and freedom, generosity and perceptions of corruption, and freedom and generosity. We call this new set of structural models model2 and apply it to the dataset in the cfa() function naming it cfa.model2. The p value of the chi-square test for this set of structural equations is 0.488.

```
model2 <- '
  national =~ GDP.per.capita +
    Social.support + Healthy.life.expectancy

  personal =~ Freedom.to.make.life.choices +
    Generosity + Perceptions.of.corruption

  personal~~national

  Social.support ~~ Freedom.to.make.life.choices

  Generosity ~~ Perceptions.of.corruption

  Freedom.to.make.life.choices ~~ Generosity
'

cfa.model2 <- cfa(model2,
    sample.cov=WorldHappinessCov,
    sample.nobs = 156, std.lv = TRUE)

summary(cfa.model2)
```

```
Model Test User Model:

  Test statistic                              4.441
  Degrees of freedom                              5
  P-value (Chi-square)                        0.488

Latent variables:
                   Estimate  Std.Err  z-value  P(>|z|)
  national =~
    GDP.per.capita    0.374    0.025   15.063    0.000
    Social.support    0.239    0.020   11.815    0.000
    Hlthy.lf.xpctn    0.215    0.016   13.777    0.000
  personal =~
    Frdm.t.mk.lf.c    0.112    0.016    6.978    0.000
    Generosity       -0.012    0.016   -0.765    0.444
    Prcptns.f.crrp    0.056    0.009    5.885    0.000
```

If we apply the lavaanPlot() function to cfa.model2 to get the path diagram, we can see that although the chi-square p value indicates that the structural equations fits the dataset, the observed variable Generosity's factor loading is shown to be insignificant; its p value is 0.444.

```
library(lavaanPlot)
lavaanPlot(model = cfa.model2,
    node_options = list(shape = "box",
    fontname = "Helvetica"),
    edge_options = list(color = "grey"),
    coefs = TRUE, covs=TRUE,
    stars = c("latent", "covs"))
```

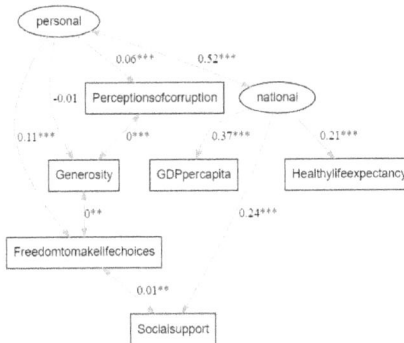

To improve the structural equations' fit with the dataset, we can remove the insignificant observed variable Generosity and any of its correlations. We can call this new set of structural equations model3 and perform CFA with it, naming it cfa.model3.

```
model3 <- '
    national =~ GDP.per.capita
    + Social.support + Healthy.life.expectancy

    personal =~ Freedom.to.make.life.choices +
    Perceptions.of.corruption

    personal~~national

    Social.support ~~
        Freedom.to.make.life.choices
    '

cfa.model3 <- cfa(model3,
    sample.cov=WorldHappinessCov,
    sample.nobs = 156, std.lv = TRUE)

summary(cfa.model3)
```

```
Model Test User Model:

    Test statistic                          3.136
    Degrees of freedom                          3
    P-value (Chi-square)                    0.371
```

We can see from the summary output of cfa.model3 that the chi-square p value is 0.371, which makes model3 a good fit for the dataset. Also from the Latent Variables and Covariances sections of the summary output as well as from the path diagram, we can see that the latent variables are all significantly inferred from all of their observed variables.

```
Latent Variables:
                    Estimate  Std.Err  z-value  P(>|z|)
  national =~
    GDP.per.capita    0.373    0.025   14.979    0.000
    Social.support    0.239    0.020   11.843    0.000
    Hlthy.lf.xpctn    0.215    0.016   13.823    0.000
  personal =~
    Frdm.t.mk.lf.c    0.112    0.016    6.969    0.000
    Prcptns.f.crrp    0.056    0.009    5.885    0.000

Covariances:
                    Estimate  Std.Err  z-value  P(>|z|)
  national ~~
    personal          0.522    0.085    6.127    0.000
 .Social.support ~~
   .Frdm.t.mk.lf.c    0.006    0.002    3.108    0.002
```

```
lavaanPlot(model = cfa.model3,
    node_options = list(shape = "box",
    fontname = "Helvetica"),
    edge_options = list(color = "grey"),
    coefs = TRUE, covs=TRUE,
    stars = c("latent", "covs"))
```

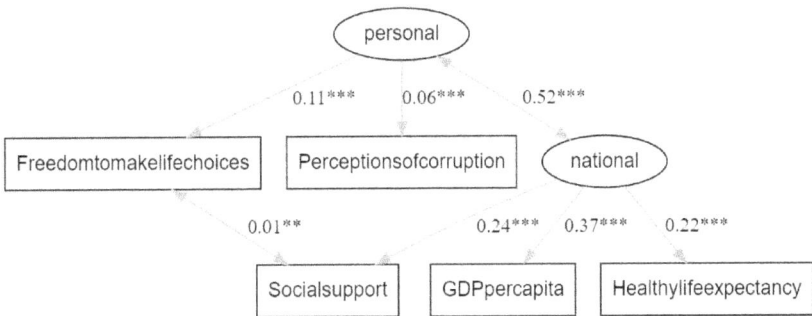

Now that you have learned how to perform a confirmatory factor analysis and how to find the best structural models for inferring latent variables, it is suggested as practice that you find a different dataset, theorize your own models, and tweak the models to best fit the dataset.

Multiple Groups CFA

Sometimes, a dataset may be divided into two or more groups of individuals who may require different sets of structural equations. In this scenario, Multiple Groups CFA may be used, as it was designed for comparing whether the structure equation models are the same between groups in a dataset. It involves testing whether the factor loadings of the observed variables of the structural equations are similar or different enough between the groups to deem the structural equations as the same or different. It involves computing the information for the groups separately, and then conducting a chi-square test to test the significance of the difference. A dataset where Multiple Groups CFA can be applied is the Mental Health in Tech Survey (OSMI, 2016). In this example, we will be comparing mental health resource access and current mental health condition of workers in the IT industry, broken down between those working remotely and those who do not work remotely.

The first step, as in basic CFA, is to load the dataset into R and assign a dataset name, here MentalHealthTech2016. Then we split the dataset into people who work remotely and people who do not work remotely using the *Remote-ORNot* variable in the dataset. Because RemoteORNot is a string variable with 2 possible entries "Yes" or "No", we can use the square bracket method to slice the dataset into people who have "Yes" in the RemoteORNot variable (assign it to the name *Remote*) and the people who have "No" (assign it to the name *NotRemote*).

```
library(lavaan)
MentalHealthTech2016 <-
    read.csv("MentalHealthTech2016.csv")

Remote <-
    MentalHealthTech2016[MentalHealthTech2016
    $RemoteORNot=="Yes",]
NotRemote <-
    MentalHealthTech2016[MentalHealthTech2016
    $RemoteORNot=="No",]
```

We then perform the CFA analysis of remote workers and non-remote workers separately:

```
MHInTechRemoteCov <- cov(Remote[1:6])
MHInTechNotRemoteCov <- cov(NotRemote[1:6])
```

First, we take the variance-covariance matrices of the remote and non-remote workers, using all columns except the one which labels each worker as remote and non-remote. This means using the first 6 columns for each of the Remote and NotRemote groups and assigning them to names *MHINTechRemoteCov* and *MHInTechNotRemoteCov* respectively.

For the structural equations of the two datasets in this example, we theorize that the latent variable *access to mental health coverage* may be inferred from the observed variables company size, mental health coverage by the company, and anonymity of using mental health resources; while the latent variable *current employee mental health condition* may be inferred from family mental health history, mental illness diagnosis, and knowledge of company coverage options. We also theorize that the access to mental health coverage and the current mental health condition of employees are correlated. A basic CFA analysis is conducted on each of the groups, remote and non-remote workers.

```
# For Remote Workers
model1.remote <- '
  MedAccess =~ CompanySize +
    MHCompanyCoverage + Anonymity

  CurrentMHCondition =~ MHFamilyHistory
    + MIDiagnoses
    + CompanyCoverageOptionKnowledge

  MedAccess~~CurrentMHCondition
'
cfa.model1.remote <- cfa(model1.remote,
    sample.cov=as.matrix(MHInTechRemoteCov),
    sample.nobs = 217, std.lv = TRUE)

summary(cfa.model1.remote)
modindices(cfa.model1.remote)
```

(Examine the output yourself and, if desired, calculate the necessary adjustments.)

For the remote workers, the final structural equations that were found to fit the variance-covariance matrix of their data involved the initial theorized equations as well as correlation equations of family history and diagnosis of mental illness, and company size with company coverage. We name this model *model2.remote*.

```
model2.remote <- '
  MedAccess =~ CompanySize
    + MHCompanyCoverage + Anonymity

  CurrentMHCondition =~ MHFamilyHistory +
    MIDiagnoses + CompanyCoverageOptionKnowledge

  MedAccess~~CurrentMHCondition

  MHFamilyHistory~~MIDiagnoses

  CompanySize~~MHCompanyCoverage
  '

cfa.model2.remote <- cfa(model2.remote ,
    sample.cov=as.matrix(MHInTechRemoteCov),
    sample.nobs = 217, std.lv = TRUE)

summary(cfa.model2.remote)
```

```
lavaan 0.6-9 ended normally after 43 iterations

  Estimator                                      ML
  Optimization method                        NLMINB
  Number of model parameters                     15

  Number of observations                        217

Model Test User Model:

  Test statistic                             11.572
  Degrees of freedom                              6
  P-value (Chi-square)                        0.072


Latent Variables:
                    Estimate  Std.Err  z-value  P(>|z|)
  MedAccess =~
    CompanySize        0.277    0.126    2.204    0.028
    MHCompanyCovrg     0.358    0.044    8.185    0.000
    Anonymity          0.208    0.034    6.064    0.000
  CurrentMHCondition =~
    MHFamilyHistry     0.121    0.040    3.050    0.002
    MIDiagnoses        0.106    0.040    2.688    0.007
    CmpnyCvrgOptnK     0.336    0.057    5.878    0.000

Covariances:
                    Estimate  Std.Err  z-value  P(>|z|)
  MedAccess ~~
    CurrentMHCndtn     0.968    0.164    5.895    0.000
 .MHFamilyHistory ~~
   .MIDiagnoses        0.078    0.017    4.483    0.000
 .CompanySize ~~
   .MHCompanyCovrg     0.168    0.049    3.459    0.001
```

For the non-remote workers, we begin the analysis with the same structural equations as the remote workers:

```
model1.not.remote <- '
  MedAccess =~ CompanySize + MHCompanyCoverage
    + Anonymity
  CurrentMHCondition =~ MHFamilyHistory +
    MIDiagnoses + CompanyCoverageOptionKnowledge
  MedAccess~~CurrentMHCondition
  '
```

```
cfa.model1.not.remote <- cfa(model1.not.remote,
    sample.cov=MHInTechNotRemoteCov,
    sample.nobs = 318, std.lv=TRUE)

summary(cfa.model1.not.remote)
modindices(cfa.model1.not.remote)
```

But as we continue with the CFA and modification of the structural equations to fit them with the dataset of the non-remote workers, the structural equations of the two groups are no longer the same. For the non-remote workers, the final structural equations are different from remote workers because the additional correlations they have are rather different, including that of anonymity with knowledge of mental health coverage options.

```
model3.not.remote <- '
  MedAccess =~ CompanySize +
     MHCompanyCoverage + Anonymity
  CurrentMHCondition =~ MHFamilyHistory
     + MIDiagnoses
     + CompanyCoverageOptionKnowledge
  MedAccess ~~ CurrentMHCondition
  MHCompanyCoverage ~~
     CompanyCoverageOptionKnowledge
  Anonymity ~~ CompanyCoverageOptionKnowledge
  '
cfa.model3.not.remote <- cfa(model3.not.remote,
    sample.cov= MHInTechNotRemoteCov,
    sample.nobs = 318, std.lv=TRUE)
```

```
summary(cfa.model3.not.remote)
```

```
lavaan 0.6-9 ended normally after 54 iterations

    Estimator                                         ML
    Optimization method                           NLMINB
    Number of model parameters                        15

    Number of observations                           318

Model Test User Model:

    Test statistic                                10.988
    Degrees of freedom                                 6
    P-value (Chi-square)                           0.089


Latent Variables:
                       Estimate   Std.Err   z-value   P(>|z|)
    MedAccess =~
        CompanySize       0.259     0.150     1.734     0.083
        MHCompanyCovrg    0.879     0.455     1.931     0.053
        Anonymity         0.081     0.048     1.710     0.087
    CurrentMHCondition =~
        MHFamilyHistry    0.328     0.049     6.706     0.000
        MIDiagnoses       0.276     0.044     6.298     0.000
        CmpnyCvrgOptnK    0.135     0.032     4.172     0.000

Covariances:
                       Estimate   Std.Err   z-value   P(>|z|)
    MedAccess ~~
        CurrentMHCndtn    0.199     0.120     1.656     0.098
   .MHCompanyCoverage ~~
       .CmpnyCvrgOptnK    0.083     0.013     6.449     0.000
   .Anonymity ~~
       .CmpnyCvrgOptnK    0.071     0.012     6.109     0.000
```

Therefore, without even considering statistical significance, we know that the structural equations for remote and non-remote workers will be significantly different. But just for the sake of demonstration, we can perform a chi-square difference test by using the anova() function on the results of the final two sets of structural equation models of the remote and non-remote workers. The p value of this test is 0 in the output, which means it is a value with many zeros after the decimal, which in turn means that we can reject the null hypothesis that the structural equations of the two datasets are the same.

```
# Chi-Square Difference Tests
anova(cfa.model3.not.remote, cfa.model2.remote)
```

```
Chi-Squared Difference Test

                      Df  AIC  BIC Chisq Chisq diff Df diff Pr(>Chisq)
cfa.model3.not.remote  6 3053 3110  11.0
cfa.model2.remote      6 2095 2146  11.6       0.583          0
```

Moving away briefly from 'Multiple Groups' CFA, the ANOVA method shown above is generally of more use when applied to the examination of models' performance over *different* datasets, when a more quantitative approach would make more sense. As well as comparative measures such as AIC and BIC, where lower scores indicate greater parsimony of fit, the fitMeasures() function includes statistics such as the NNFI (also known as the Tucker Lewis Index), where a higher score is desirable, and RMSEA.

One example, showing a proliferation of statistics, would be this default:

```
fitMeasures(cfa.model2.remote)
```

To avoid a long read-out, specify a statistic like so:

```
fitMeasures(cfa.model2.remote, "nnfi")
```

or specify a set of statistics, for example:

```
fitMeasures(cfa.model2.remote,
    c("nnfi", "rmsea", "bic", "aic"))
```

Returning to the use of ANOVA specific to multi-group CFA: It is used to highlight the differences between the factor loadings of the CFA models for each different group within the same dataset. It is acceptable to examine the groups as if they were different datasets.

Moving back to our multiple groups example: If you would like to see how different the CFA models of the remote and non-remote workers are from each other, use the lavaanPlot() function from the lavaanPlot package in the same way as in CFA, but applied to each group.

```
library(lavaanPlot)
lavaanPlot(model = cfa.model2.remote,
    node_options = list(shape = "box",
    fontname = "Helvetica"),
    edge_options = list(color = "grey"),
    coefs = TRUE, covs=TRUE,
    stars = c("latent", "covs"))
```

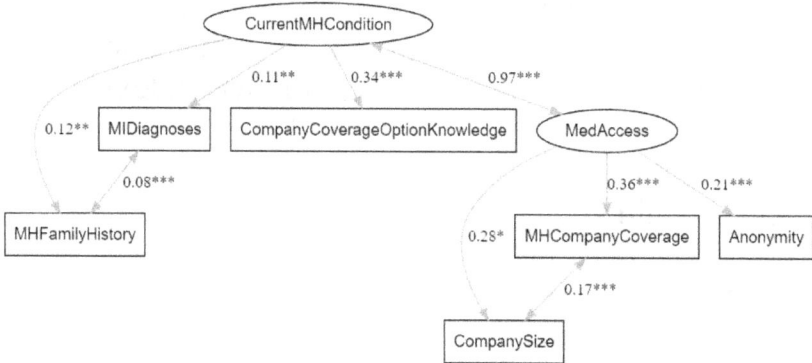

For the remote workers, we can see that the latent variables are inferred by all observed variables with statistical significance, with a significant correlation between the two latent variables CurrentMHCondition and MedAccess. From these results, and given that the chi-square test's p value in the Model Test User Model is higher than 0.05, we can say that the CFA model fits the dataset. Also, the fact that we only had to add correlations to make the CFA model fit the data – instead of having to make radical changes to the equations – indicates that we can accept that the *model2.remote* CFA model is valid for the remote workers' group.

```
lavaanPlot(model = cfa.model3.not.remote,
    node_options = list(shape = "box",
    fontname = "Helvetica"),
    edge_options = list(color = "grey"),
    coefs = TRUE,
    covs=TRUE, stars = c("latent", "covs"))
```

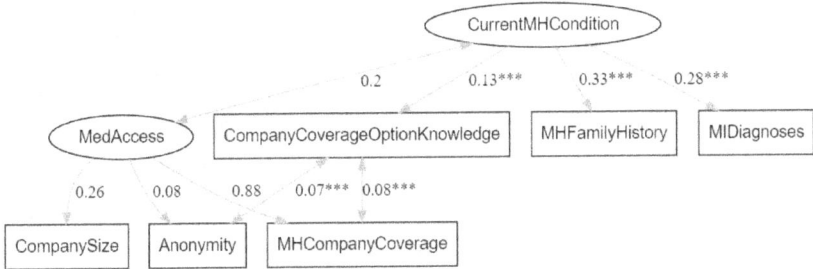

But the results of how well the theorized model fits for non-remote workers are quite different. Even after taking additional correlations into account and the Model Test User Model's chi-square test showing that the CFA model is a good fit for the dataset, we can see that the MedAccess latent variable was not significantly inferred from the observed variables. Also, the correlation between CurrentMHCondition and MedAccess is not a statistically significant one. These results suggest that *model3.not.remote* is probably not the optimum model for the non-remote workers, and that the remote and non-remote workers do not share the same CFA model. To practice this method for yourself, you could theorize other structural equations for remote or non-remote workers.

Basic SEM

Now that we have finally reached the method which shares the same name as the umbrella term of structural equation modelling, you may be wondering what path analysis and CFA have to do with SEM, and what is the purpose of SEM? Path analysis is used to determine existing causal relationships between variables in a dataset while CFA is used to see if the variables in the dataset are good for inferring latent variables we need for specific studies. For SEM, the method builds on both methods, in the sense that it focuses on testing for a significant relationship between latent variables inferred by the observable variables in the dataset. In order to accurately perform SEM, the assumption of multivariate normality should be met, observed variables need to be continuous, and the sample size should be at least 5-10 individuals for each observed variable involved in the structural equation (Klem, 2000).

The following example uses the *Bollen* dataset (Rosseel, 2012). The dataset contains multiple observed variables which can be used to infer the industrialization and political democracy of developing countries. If you look at this dataset, you may notice that it is quite small: it is used for demonstration purposes only. When you are analysing your own datasets, please try to ensure that the sample size is larger than 200 individuals (Klem, 2000).

In this example, we theorized that two latent variables, PoliDemocracy (political democracy) and Industrialization, may be inferred by the observed variables in the dataset. For PoliDemocracy, the relevant observed variables are PressFreedom (Freedom of the press), PoliOppFreedom (Freedom of Political Opposition), FairElect (Fair Elections), and LawEffective (Effectiveness of the Legislature). For Industrialization, these are GNP.per.capita (Gross National Product), EnergyConsumption.per.capita, PercentLabour-Force (percentage of the labour force in industry), and PoliDemocracy - we also theorize that industrialization has a causal relationship with political democracy.

To begin testing these theories with basic SEM, we open the lavaan package using the library() function and open the Bollen dataset. For this analysis, we are only going to use variable columns 1 to 4 and 9 to 11 in the dataset. Therefore, we take out these columns using the square bracket methods and selected the columns using a vector value of c(1:4, 9:11) and assign the data frame of these vectors with the name *DemocracyIndustry*.

```
library(lavaan)
Bollen <- read.csv("Bollen.csv")
DemocracyIndustry = Bollen[,c(1:4, 9:11)]
```

After that, we rename the columns of the DemocracyIndustry data frame by placing it in the *names*() function and assigning to it a vector of strings (names for each column):

```
names(DemocracyIndustry) <-
  c("PressFreedom", "PoliOppFreedom",
    "FairElect", "LawEffective",
    "GNP.per.capita",
    "EnergyConsumption.per.capita",
    "PercentLabourForce")

CovDemocracyIndustry <- cov(DemocracyIndustry)
```

As with CFA, we then convert this data frame into a variance-covariance matrix using the *cov*() function and assign it the name CovDemocracyIndustry.

To perform the SEM test for the structural equations, as with all methods under the structural equation modeling umbrella, we need to first specify the structural equations. Because PoliDemocracy and Industrialization are latent variables, they are specified as dependent variables with =~ between them and their individual observable variables. Also, we want to test if Industrialization has a causal relationship with Polidemocracy, so we put PoliDemocracy as a dependent variable and Industrialization as the independent variable with a tilde (~) between them.

```
basic.model1 <- '
  PoliDemocracy =~ PressFreedom +
    PoliOppFreedom + FairElect + LawEffective
```

```
Industrialization =~ GNP.per.capita +
  EnergyConsumption.per.capita +
  PercentLabourForce

PoliDemocracy ~ Industrialization
  '
```

After the structural equations are specified, we use the *sem*() function from the lavaan package to test if these structural equations fit the variance-covariance matrix of the dataset (and in turn the dataset itself, DemocracyIndustry). We take the name of the structural equations object (which was assigned the name *basic.model1*) and put it as the first entry of the sem() function. We then put the variance-covariance matrix of DemocracyIndustry (CovDemocracyIndustry) as the second entry and the number of individuals in the dataset as the third entry. We then assign this entire function the name *basic.fit1*:

```
basic.fit1 <- sem(model=basic.model1,
    sample.cov = CovDemocracyIndustry,
    sample.nobs = 75, std.lv=TRUE)
summary(basic.fit1)
```

If we look at the outcome of the test by applying the summary() function to basic.fit1, we can see that the structural equations we specified do not fit the variance-covariance matrix of this dataset significantly because the p value of the chi-square test is 0.034, lower than 0.05.

```
lavaan 0.6-9 ended normally after 25 iterations

    Estimator                                     ML
    Optimization method                       NLMINB
    Number of model parameters                    15

    Number of observations                        75

Model Test User Model:

    Test statistic                            23.738
    Degrees of freedom                            13
    P-value (Chi-square)                       0.034
```

```
modindices(basic.fit1)
```

As in path analysis and CFA, we use modification indices to modify our structural equations by finding and taking into consideration one or more correlation effects in the structural equations. In this case, the correlation we found using the modindices() function is the correlation between PressFreedom and FairElect. By including this correlation in the structural equations and rerunning the test, we can see that this new structural equation fits the variance-covariance matrix significantly, indicated by a chi-square test p value of 0.300.

```
basic.model2 <- '
  PoliDemocracy =~
      PressFreedom + PoliOppFreedom +
      FairElect + LawEffective
  Industrialization =~ GNP.per.capita +
      EnergyConsumption.per.capita +
      PercentLabourForce
  PoliDemocracy ~ Industrialization
  PressFreedom ~~ FairElect
  '
```

```
basic.fit2 <- sem(model=basic.model2,
    sample.cov = CovDemocracyIndustry,
    sample.nobs = 75, std.lv=TRUE)
summary(basic.fit2)
```

```
Model Test User Model:

  Test statistic                        14.012
  Degrees of freedom                        12
  P-value (Chi-square)                   0.300


Latent Variables:
                   Estimate  Std.Err  z-value  P(>|z|)
  PoliDemocracy =~
    PressFreedom      1.682    0.239    7.027    0.000
    PoliOppFreedom    2.596    0.357    7.269    0.000
    FairElect         1.819    0.312    5.839    0.000
    LawEffective      2.810    0.293    9.594    0.000
  Industrialization =~
    GNP.per.capita    0.670    0.065   10.349    0.000
    EnrgyCnsmptn..    1.459    0.128   11.399    0.000
    PercentLabrFrc    1.218    0.128    9.486    0.000

Regressions:
                   Estimate  Std.Err  z-value  P(>|z|)
  PoliDemocracy ~
    Industrializtn    0.534    0.142    3.771    0.000

Covariances:
                   Estimate  Std.Err  z-value  P(>|z|)
 .PressFreedom ~~
   .FairElect         1.826    0.659    2.771    0.006
```

```
library(lavaanPlot)
lavaanPlot(model = basic.fit2,
  node_options =
  list(shape = "box", fontname = "Helvetica"),
  edge_options = list(color = "grey"),
  coefs = TRUE,
  covs=TRUE, stars = c("regress", "latent"))
```

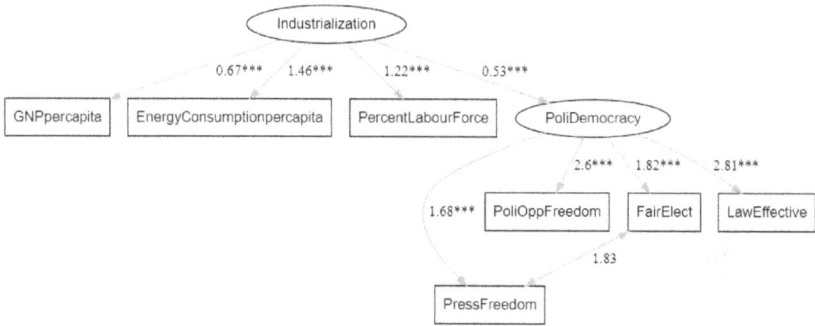

If we take a look at the output information and path diagram of basic.fit2, we can see that the latent variables were significantly inferred by all of the relevant observed variables. Also, the latent variable Industrialization had a significant influence on PoliDemocracy. If you want to practice this method further, there are many other datasets in the lavaan package. You may also apply SEM to the example datasets used for path analysis and CFA to infer other latent variables and see if they have significant relationships with each other.

Latent Growth Models

Latent growth models (aka longitudinal growth models or longitudinal SEM models) analyze repeated measures (longitudinal data), although they also fall under the Structural Equation Modeling umbrella. When we analyze longitudinal data, our main interest is usually the change of one or more dependent variables over time, and we focus on the changes of the intercepts (means) and slopes (rate of change) of the variable overtime. Latent Growth Models can be created from longitudinal datasets that meet the assumption of multivariate normality and have a large sample size. The most accurate results are achieved if there is the same length of time between each time interval (Bollen and Curran, 2006).

For Latent Growth Models, the dependent variable is a latent variable. We look at the changes in intercepts and slopes (means and rate of change) while testing the significance of the possible predictor or other covariate variables that could influence the dependent variable. The following example is a simulated dataset of 500 American millennial young professionals aged 22 – 26 who are in long term relationships, showing percentage changes in their motivation to have biological children. (The data was simulated using the variance-covariance matrix of another example dataset, *Demo.growth*, in the lavaan package (Rosseel, 2012). The predictors are the career motivation of the individuals and their spouses' desire for biological children; the time-varying covariate is the individuals' parents' desire for biological grandchildren at each time-point.

Growth Latent Modeling can be performed with the lavaan package. The first step is to open the package in the R environment using the library function and read the dataset into the R environment using the read.csv() function.

```
library(lavaan)
ChildDesire4Years <-
    read.csv("ChildDesire4Years.csv")
```

To perform Growth Latent Models in R, it is required that the dependent variable(s) and time-varying covariates should have their own columns at each time-point. In this example, the dependent variable (the desire to have biological children, which is time-varying) and time-varying covariate (parents' desire to have biological grandchildren) each have 4 columns in the dataset (one for each timepoint): ChildDesire1, ChildDesire2, ChildDesire3, ChildDesire4, and ParentsDesire1, ParentsDesire2, ParentsDesire3, ParentsDesire4. The two time-invariant covariates CareerMotivation and SpouseChildDesire each have 1 column. Here is a sample of the data:

```
ChildDesire4Years[1:2, 1:5 ]
```

	ChildDesire1	ChildDesire2	ChildDesire3	ChildDesire4	CareerMotivation
1	0.01363084	0.7529748	1.8353802	1.206944	1.109702
2	-0.65859912	1.5201076	0.9975355	2.617161	0.797728

```
ChildDesire4Years[1:2, 6:10 ]
```

	SpouseChildDesire	ParentsDesire1	ParentsDesire2	ParentsDesire3	ParentsDesire4
1	-0.2600287	0.67062195	-0.91057389	-0.9266877	-1.7206268
2	-1.2384914	0.05253744	-0.02541313	-1.2584188	0.7079315

Like all other methods that fall under the SEM umbrella, we need to specify the structural equations before testing to see if the equations fit the variance-covariance matrix of the dataset. Specifying the structural equations of a longitudinal dataset is slightly more complicated because the dataset involves repeated measures. The first step is to specify equations that infer the intercept (represented by i) and slope (represented by s) as dependent variables on the left side of the `=~` sign while the ChildDesire variables over the four timepoints are used as the independent variables. For each independent variable, there is a number followed by an asterisk (*) in front of them. The asterisk symbolizes multiplication and the numbers are the coefficients of the equations.

```
growth.model <- '
i =~ 1*ChildDesire1 + 1*ChildDesire2 +
     1*ChildDesire3 + 1*ChildDesire4
s =~ 0*ChildDesire1 + 1*ChildDesire2 +
     2*ChildDesire3 + 3*ChildDesire4

i+s ~ CareerMotivation + SpouseChildDesire

ChildDesire1 ~ ParentsDesire1
ChildDesire2 ~ ParentsDesire2
ChildDesire3 ~ ParentsDesire3
ChildDesire4 ~ ParentsDesire4
  '
```

The reason that all coefficients for the intercept equation are 1's is because the intercept needs the coefficients to be fixed; the reason why the coefficients for the s (slope) equation are 0, 1, 2, and 3 is to establish a linear trend for the slopes at each timepoint with 0 used as the starting point. (N.b. Please note that the s coefficients are not set as actual coefficients for the model. The real coefficients calculated using the dataset will be shown in the output.)

CareerMotivation and SpouseChildDesire are predictor variables which are suspected to influence the desire for children, which means we suspect they will influence the intercept and the slope of the model. Therefore, we add another equation which represents this relationship where i+s is on the left side of ~ as dependent variables while CareerMotivation + SpouseChildDesire is on the right representing the independent variables.

Next, since the parents' desire for biological grandchildren is a time-varying covariate, which means that the variable at each timepoint can only influence each young professional's desire to have biological children at their own timepoints, we can put down four equations for ParentsDesire's influence on ChildDesire for each timepoint. We assign the string which contains all seven equations the name *growth.model.*

To fit the Growth Latent Model, we use the *growth*() function from lavaan. The first entry is the structural equation we specified, growth.model, and the second entry is the name of the dataset ChildDesire4Years. We assign the entire model the name growth.fit. To examine the model, we apply the summary() function to growth.fit.

```
growth.fit = growth(growth.model,
          data=ChildDesire4Years)
summary(growth.fit)
```

The way of concluding whether or not a Growth Latent Model fits the dataset is the converse of path analysis, CFA, and SEM. For path analysis, CFA, and SEM, the p

value of the chi-square test should be above 0.05 while for Growth Latent Model the p value should be below 0.05. In the output of growth.model, the p value is 0.046, hence the model we specified fits the ChildDesire4Years dataset significantly.

```
lavaan 0.6-9 ended normally after 31 iterations

  Estimator                                      ML
  Optimization method                        NLMINB
  Number of model parameters                     17

  Number of observations                        500

Model Test User Model:

  Test statistic                             33.016
  Degrees of freedom                             21
  P-value (Chi-square)                        0.046


Latent Variables:
                   Estimate  Std.Err  z-value  P(>|z|)
  i =~
    ChildDesire1      1.000
    ChildDesire2      1.000
    ChildDesire3      1.000
    ChildDesire4      1.000
  s =~
    ChildDesire1      0.000
    ChildDesire2      1.000
    ChildDesire3      2.000
    ChildDesire4      3.000
```

Also, if we take a look at the regressions section of the output, we can see that for both intercept and slope, predictors CareerMotivation and SpouseChildDesire have significant influence on them; and the time-varying covariates ParentsDesire for each year had a significant influence on ChildDesire for each year:

```
Regressions:
                    Estimate  Std.Err   z-value  P(>|z|)
  i ~
    CareerMotivatn    0.608     0.054    11.330    0.000
    SpouseChildDsr    0.612     0.058    10.522    0.000
  s ~
    CareerMotivatn    0.262     0.025    10.282    0.000
    SpouseChildDsr    0.529     0.028    19.101    0.000
  ChildDesire1 ~
    ParentsDesire1    0.146     0.045     3.224    0.001
  ChildDesire2 ~
    ParentsDesire2    0.289     0.041     7.037    0.000
  ChildDesire3 ~
    ParentsDesire3    0.324     0.040     8.203    0.000
  ChildDesire4 ~
    ParentsDesire4    0.339     0.054     6.316    0.000
```

This example is straightforward in the sense that the structural equations we specified fit the dataset significantly without the need to make changes to it. Now it's your turn to find a longitudinal dataset of your own and specify your own structural equations to see if they fit the dataset. It is possible that the structural equations can fit the dataset with statistical significance while the predictors may not significantly influence the latent dependent variable. In cases like this, it's best to remove the insignificant predictor variable and fit the model to the dataset again.

Further reading

If you would like to read more about the underlying theories of the methods presented in the SEM section, there are many published books and articles in statistical journals that can satisfy your curiosity. For path analysis, CFA, and SEM, I recommend *Beginner's Guide to Structural Equation Modelling* by Schumacker and Lomax (2010). For latent growth models, I recommend *Latent Growth Models: A Structural Equation Perspective* by Bollen and Curran (2006). Both books offer in-depth explanations of the theory behind the methods.

Chapter 2 – Time Series Analysis

Sections include

Stationary data vs. non-stationary data

Making non-stationary data stationary

ARIMA

Auto ARIMA

Seasonal ARIMA (SARIMA)

Introduction

Time series analysis is one of the most widely used areas of statistical analysis for forecasting outcomes of situations close to our daily lives, such as stock prices and sales, weather temperature, household energy consumption, etc. These datasets are called time series data because the data of a single variable are collected in equally spaced time-points over a period of time. It is assumed that the data of the variable contains trends where future trends can be predicted using its past trends. Other than trends, it is also assumed that most time series datasets have some form of seasonality where the values of the variable at a certain time are higher/lower than other times (e.g. the weather temperature is higher during summer than the other three seasons). Before we get into the specifics of the methods used to analyze time series data, I would like to mention that the focus of this chapter is on linear, univariate time series analysis. Non-linear and multivariate time series analysis are beyond the scope of this book.

The best way to understand time series analysis is by comparing it with the simplest statistical model, linear regression. The purpose of linear regression is to model the influence of an independent variable (or variables) on an outcome variable.

Outcome =
  Coefficient x Independent + NoiseOfOutcome

To illustrate this, we have the outcome variable on the left-hand side of an = sign while on the right-hand side of the equal sign, we have the independent variables and their coefficients (which represent the degrees and directions of influence the independent variables have on the outcome variable). Also included on the right-hand side of the = sign are the parts of the outcome variable that are not influenced by the independent variable, which we call here NoiseOfOutcome.

Unlike linear regression, the outcome and independent variables in a time series model are not separate variables – they are the values of the same variable at different timepoints. The model represents the influence of the value of the outcome variable at a previous timepoint on the value of the outcome variable at the current timepoint.

CurrentOutcome =
    Coefficient x PrevTimepointOutcome
    + CurrentNoiseOfOutcome

The degree and direction of the value of the previous time-point's influence are represented by the coefficient. The noise (at the current timepoint) represents the parts of the current value of the outcome variable that are not influenced by the value of the previous time point. Because in time series analysis, the model represents the outcome variable's regression on itself, a time-series model is called an **autoregression** (AR or self-regression) model.

In time-series analysis, we assume that the values of the outcome variable at each time point correlate with each other. Therefore, it is correct to assume that the noise of the outcome variable value at each timepoint also correlates. Therefore, the current noise of the outcome variable is equal to the noise of the outcome variable value at the previous timepoint with its coefficient, plus the noise of the current noise of the outcome variable value. This model is called the **moving average** (MA) model:

CurrentNoiseOfOutcome =
   Coefficient x NoiseofOutcomePrevTP
   + **NoiseOfCurrentNoiseOfOutcome**

The model we use to model time series data is a combination of the autoregression model (AR) and the moving average model (MA) called the **ARIMA** (AutoRegressive Integrated Moving Average) model, which models the current outcome variable value using the outcome value of the previous timepoint and the noise of the outcome of the previous timepoint (Shumway and Stoffer, 2017).

CurrentOutcome =
   Coefficient x PrevTimepointOutcome
   + **Coefficient x NoiseofOutcomePrevTP**
   + **NoiseOfCurrentNoiseOfOutcome**

Stationary Data vs. Non-Stationary Data

The ARIMA model seems like a great method for modelling time series data, but there is a catch! Like all statistical models and tests, there are conditions for time series datasets that must be met. In order to be deemed suitable for ARIMA, the time series data needs to be deemed **stationary** and for a time series data to be deemed stationary, it must meet three criteria:

1) The time series must have **constant mean** (which means there is no trend in the data)

2) The time series must have **constant variance** (in other words, the amplitude values of the variable at different time points should not differ too much)

3) The time series must have an **autocovariance that does not depend on time** (the frequency of the changes in values should not differ as time passes) (Box *et al*, 1994).

To illustrate the three criteria graphically, the four graphs below show stationary, non-constant mean, non-constant variance, and time dependent auto-covariance in time series data. The wavy lines in each graph represent the values of the variable in question over time; the straight line represents the mean values of the variable, for each time frame, over time.

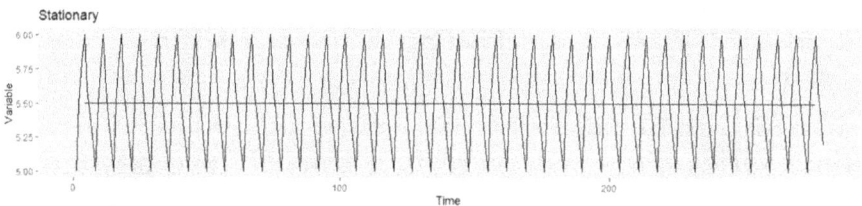

For the stationary data, we can see that the mean line is straight (which shows that the means between time frames are constant) and the peaks and valleys of the graph are relatively similar to each other between time frames (which shows that the variance is constant and the autocovariance is not time-dependent).

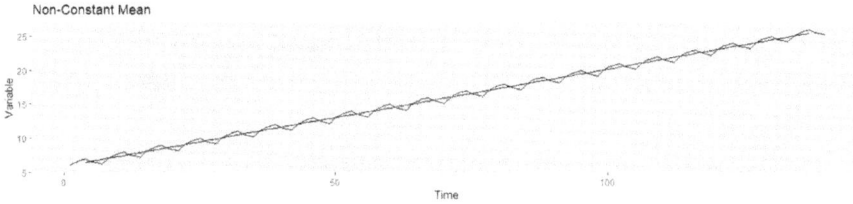

Non-Constant Mean

In the non-constant mean graph, we can see that the time series graph and the means between time frames are increasing instead of being constant as time progresses.

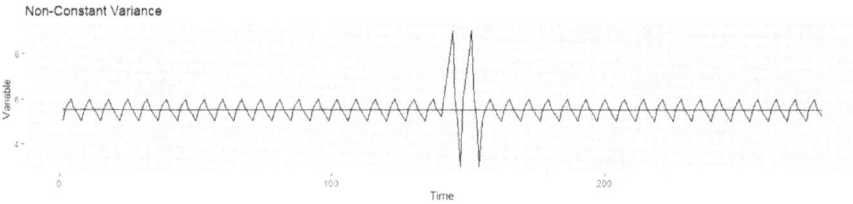

Non-Constant Variance

In the graph which shows non-constant variance, the means between time frames are constant, but non-constant variance is shown as the significantly higher amplitude of the graph between timepoints 141 and 154.

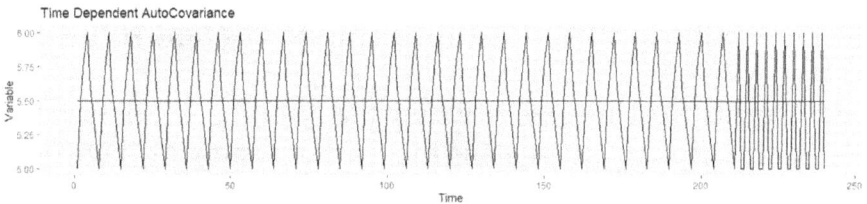

Time Dependent AutoCovariance

In the time dependent autocovariance graph, as in the non-constant variance graph, the means between time frames are constant, but the time-dependent autocovariance is shown as an increase of graph frequency (where there are more waves for each time frame) between timepoints 211 and 240 (Boashash, 2015; Hyndman and Athanasopoulos, 2018).

Although in these graphs, stationarity can be seen clearly, as can violations, in real life datasets the graphs are usually not as clear. How straight and flat does the mean line have to be in order to be considered to have constant mean? How different do the amplitude and frequency of certain sections of the time series have to be to be considered to have non-constant variance or time-dependent autocovariance? Luckily for us, there are multiple ways we can test for the stationarity of time series data statistically instead of visually.

Testing Stationarity of Data

Before we get into the tests to examine stationarity of time series data, we need to first understand that there are three types of stationarity:

1. **Strict Stationary** – when the mean, variance, and covariance of the time series data are not changing with time. It is the ideal stationarity for modelling time series data using ARIMA.

2. **Trend Stationary** – when the time series data has a trend (increasing/decreasing mean), but removing it will lead to a strict stationary dataset.

3. **Difference Stationary** – if 'detrending' does not work, the time series data can be made stationary through a process called **differencing**, where we take the difference between the variable value at a particular instant and the value at the previous instant (Box *et al*, 1994). Datasets with this kind of non-stationarity are also known to have a **random walk trend** (Hyndman and Athanasopoulos, 2018).

There are two main methods for testing for the three types of stationarity of data:

1) **Augmented Dickey-Fuller test** (aka ADF, or Unit Root test)

Null Hypothesis: The time series is not stationary (but it could be difference stationary)

Alternative Hypothesis: The time series is stationary (Greene, 2002)

2) **KPSS (Kwiatkowski-Phillips-Schmidt-Shin) test**

Null Hypothesis: The time series data is trend stationary

Alternative Hypothesis: The time series is not stationary (Kwiatkowski *et al*, 1992)

The ADF and KPSS tests are usually used together to infer the stationarity of the time series data. There are four possible combinations of the results of the two tests:

1) **Both ADF and KPSS conclude that the data is not stationary**: we can assume the data is not stationary

2) **Both ADF and KPSS conclude that the series is stationary**: we can assume the data is strictly stationary

3) **ADF concludes not stationary, KPSS concludes trend**

stationary: we have trend stationary data where we need to remove the trend to make the data strictly stationary

4) **ADF concludes stationary, KPSS concludes not trend stationary:** we have difference stationary and can use differencing to make data strictly stationary (Singh, 2018).

Making Non-Stationary Time Series Stationary

So what happens if the time series data in question is shown to be not strictly stationary? In order to be able to fit an ARIMA with non-stationary time series data, we must first make the time series data stationary. There are many methods for making non-stationary time series data stationary. Depending on the level of non-stationarity and type of non-stationarity, the methods you choose to apply when you first start your own analysis can be based on your understanding, intuition, and experience. The bottom line is that we must do due diligence to ensure that the data is as stationary as possible based on our knowledge before we apply ARIMA or any other modelling methods to the dataset.

To start with, we can look at whether the data has trends (where the mean is varied over time) or seasonality (higher values for the time series at certain periods and lower at other periods). For data that only has trends, we can start with some elementary methods to estimate the trend in the data such as **aggregation** or **smoothing** (aka rolling means).

Aggregation is performed by dividing the entire time series into equal time frames and creating a trend line using the mean value of each time frame.

Smoothing is carried out by placing ticks onto the entire time series progression to divide it into equal time frames, then take the means of n values (depending on the frequency of the time series) before each tick to create a trend line. Before estimating the trend, it is recommended that we should perform a data transformation (log, square root, etc.) to assist in decreasing the trend of the time series. After the trend has been estimated using one of these methods, we minus the trend out of the time series to try and make the non-stationary time series stationary (Shumway and Stoffer, 2017).

Although aggregation and smoothing can work when the non-stationarity is not too complicated, they may at times fail to completely remove the non-stationarity from a time series, especially when the time series data involves seasonality. In these situations, methods such as **differencing** and **decomposition** can be highly useful. Differencing involves taking the difference of two instances of the time series. Decomposing involves modelling the trend and seasonality of the time series separately (Shumway and Stoffer, 2017), removing them from data and giving you what is left over: the non-stationary data!

Whew! We have just spent several pages of the time series section talking about only the theory and the methods used in time series analysis. Some of you might be thinking "April, I thought this is supposed to be a book that teaches us how to perform statistical analyses, not just discussing theories and methods!" Although I completely understand your feelings of indignation, the reason I did not show an example until now is because as much as time series analysis is a mechanical technique in data analysis, it is also an art form. What this means is that the methods we have discussed so far can be used selectively and in no particular order (depending on the data) when performing a time series analysis. To demonstrate what I mean, without further delay, let's go through an example!

Example: Carbon dioxide in the air

For our first example, let's look at the carbon dioxide air concentration dataset from the Mauna Loa observatory in Hawaii, USA. The data were collected monthly between the years 1975 and 1986 (U.C. San Diego, 2016). We first install and open the R packages that are needed for this example, which are the packages readxl, zoo, ggplot2, tseries, forecast, and lmtest. We install them by entering the package names with quotation marks (" ") around them into the install.packages() function, and open the installed packages by entering their names without quotation marks into the library() function.

```
install.packages("readxl")
install.packages("zoo")
install.packages("ggplot2")
install.packages("tseries")
install.packages("forecast")
install.packages("lmtest")
```

```
library(readxl)
library(zoo)
library(ggplot2)
library(tseries)
library(forecast)
library(lmtest)
```

Next, we read in the dataset CO_2_Concentration into R using the read_excel() function from the readxl package, because the dataset is in the xlsx format. The dataset is assigned the name CO_2_Concentration:

```
CO2_Concentration <-
  read_excel("CO2_Concentration.xlsx")
CO2_ConcentrationCut <-
  CO2_Concentration[9:152, 1:2]
```

In this example, we are only going to use rows 9 – 152 of the dataset. For this reason, we need to cut the CO2-_Concentration dataset using the square-bracket method where only those rows and columns are selected. The rows to be kept are indicated as 9:152, meaning rows 9 to 152. The columns to be kept are indicated as 1:2, only the first two columns. This reduced dataset is given the name CO_2_ConcentrationCut and it should have two columns: CO_2 for the CO_2 concentration, and the Year&Month variable as an indicator variable that assigns a unique and sequential value for each CO_2 concentration value in the dataset.

Next, we look at the CO2 concentration data over time by plotting it in a time series graph, using the plot() function:

```
plot(CO2_ConcentrationCut$'Year&Month',
     CO2_ConcentrationCut$CO2, type="l",
     xlab = "Time", ylab="CO2 Concentration")
```

We first enter the time variable Year&Month (The $ is used to indicate that the variables after it are columns in the CO2_ConcentrationCut data frame. Year&Month is placed in apostrophes to indicate that it is a column name). Then comes the CO2 concentration using CO2_ConcentrationCut$CO2. Then we enter type = "l" to indicate that this is a line graph and we place "Time" and "CO2 Concentration" as xlab and ylab respectively to assign names to the axes that are more understandable than the names of the columns in the data frame.

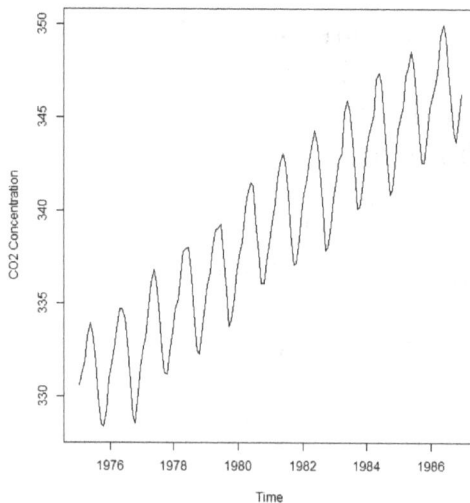

From the time series plot, we can see that the CO_2 concentration is not stationary: it is increasing over the years.

To test the stationarity of this data visually and statistically, we can plot the rolling mean and use the Dickey-Fuller Test and the KPSS test.

```
# Test for Stationarity of CO2 Data
# Make zoo object of data
CO2.zoo <- zoo(CO2_ConcentrationCut$CO2,
        CO2_ConcentrationCut$'Year&Month')
```

We can calculate the rolling mean of this dataset by using the zoo() function from the zoo package. The zoo() function is used to create a zoo object, which in this case is the CO_2 concentration values ordered by the Year&Month values. We can do this by entering CO2_ConcentrationCut$CO2 and CO2_ConcentrationCut$'Year&Month' as the first and second arguments in the zoo() function and we name this zoo object CO2.zoo. The purpose of the zoo object is to calculate the rolling mean easily using the rollmean() function in the zoo package.

```
# Calculate moving average
# and make first and last value 'NA'
# (to ensure identical length of vectors)
CO2_rollmean <- rollmean(CO2.zoo, 12,
        fill = NA, align = "right")
# Add calculated moving averages to existing
# data frame
CO2_ConcentrationCut$CO2_rollmean <-
            CO2_rollmean
```

Within the rollmean() function, the first argument is the zoo object in question, CO2.zoo. The second argument is the width of the rolling window (or the number of values that will be used to calculate each mean value). The number 12 is chosen as the width of the rolling window because this dataset was collected monthly between 1975 and 1986; so calculating a mean for every 12-monthly CO2 concentration period makes the rolling mean easy to track. The arguments fill = NA and align = "right" specify all the rolling mean values, which will be a series with 133 mean values (144 - 11 values). The rolling means series is given the name CO2_rollmean.

In the second expression, a new column is created. CO2_rollmean is placed by assignment within the CO2_ConcentrationCut data frame as a column named CO2_ConcentrationCut$CO2_rollmean.

```
p <- ggplot(CO2_ConcentrationCut,
    aes('Year&Month', CO2)) + geom_line() +
    geom_line(aes('Year&Month', CO2_rollmean),
    color="black", linewidth=2) +
    xlab("Time") + ylab("CO2")

print(p + ggtitle("CO2 Concentration in Air
    over Time"))
```

After the rolling means have been calculated, we use the functions from the ggplot2 package to plot the CO2 concentration and the rolling means together. First we use the ggplot() function, entering the data frame name as the first argument. The second argument uses the aes() function with the columns Year&Month and CO2, supplying the x and y-values for the time series plot respectively. Then we use + geom_line() to specify that the Year&Month-CO2 plot should be a linear plot.

To add another line for the rolling mean, we put another + geom_line() with the first argument containing the aes() function with Year&Month, CO2_rollmean, to specify that Year&Month and CO2_rollmean are the columns to supply the x and y-values respectively for the rolling mean plot. In addition, the color and thickness of the rolling mean plot are set as black and twice as thick as the time series plot using color="black" and linewidth=2 respectively in the geom_line() function for the rolling mean. The x and y plot axes are labelled as Time and CO2 by adding xlab("Time") and ylab("CO2"). The plot is given the title 'CO2 Concentration in Air OverTime' using the ggtitle() function.

N.B. For this plot to come out correctly, it is necessary to use this symbol - ` - to be found on the top left of your keyboard, to begin and end the two instances of Year&Month (`Year&Month`) within the aes() function.

CO2 Concentration in Air over Time

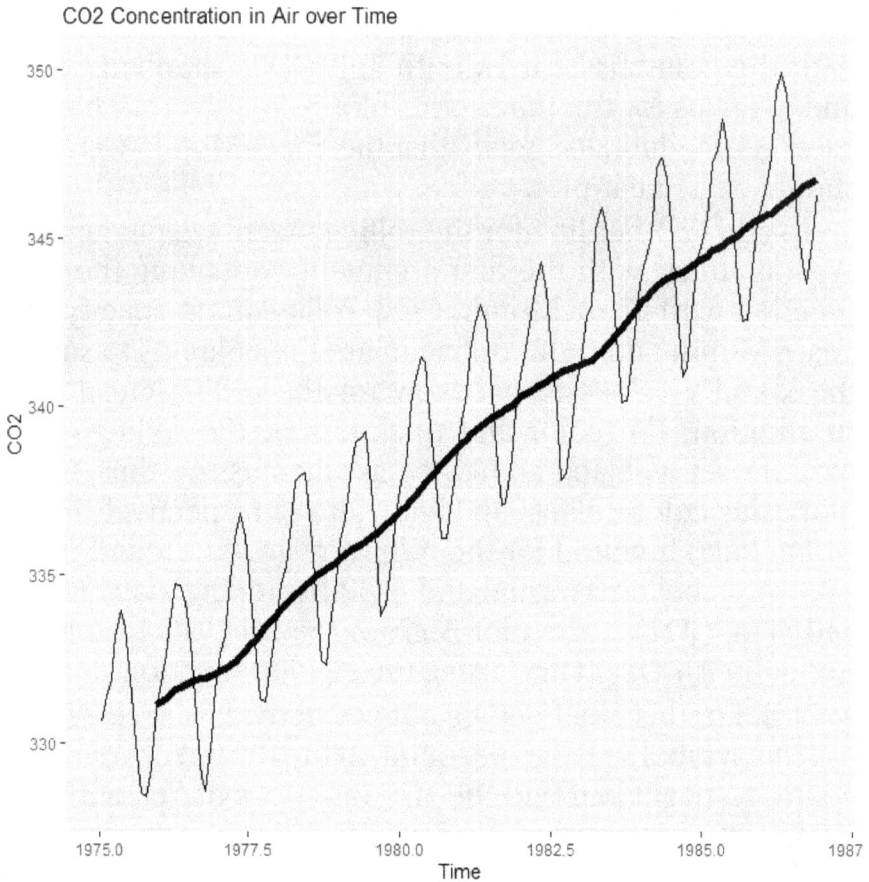

The rolling mean plot confirms our initial guess that the CO2 Concentration data is not stationary because the rolling mean is increasing instead of staying constant. To further confirm this, we use the Dickey-Fuller and KPSS tests for the CO2 concentration data. These tests can be performed using the adf.test() and kpss.test() functions from the tseries package. Both tests are performed on the actual dataset.

```
adf.test(CO2_ConcentrationCut$CO2,
     alternative= "stationary", k=12)
```

For the Dickey-Fuller test, CO2_ConcentrationCut$CO2 is the first argument for the adf.test() function and the second argument, alternative= "stationary", declares that the alternative hypothesis is that the data is stationary. The third argument states that the lag order is 12, reflecting the monthly data collection.

```
kpss.test(CO2_ConcentrationCut$CO2,
     null = "Trend", lshort=FALSE)
```

The KPSS test has the same first argument. The second argument, null = "Trend", means that the null hypothesis is that the data is trend stationary and the alternative hypothesis is that the data is not stationary. The third argument, lshort=FALSE, is set for a relatively long-term lag, as in the current case of the 12 months lag stipulated in the adf.test() function.

```
        Augmented Dickey-Fuller Test

data:  CO2_ConcentrationCut$CO2
Dickey-Fuller = -1.7909, Lag order = 12, p-value = 0.6635
alternative hypothesis: stationary
```

For ADF, with an alternative hypothesis of 'stationary' and a null hypothesis of 'non-stationary', a non-significant result - as in this case – indicates that the null hypothesis cannot be rejected, which in turn leads to non-stationary data. (In practical terms, a high p value for the ADF test indicates non-stationary data; $p < 0.05$ indicates stationary data.)

```
       KPSS Test for Trend Stationarity
data:  CO2_ConcentrationCut$CO2
KPSS Trend = 0.11355, Truncation lag parameter = 13, p-value = 0.1
```

The null hypothesis in KPSS is that the data is trend stationary while the alternative hypothesis is that the data is not trend stationary. So a non-significant result in KPSS - as in this case – means the null hypothesis cannot be rejected, which represents trend stationary data. (In practical terms, a high p value for the KPSS test indicates trend stationary; $p < .05$ indicates non-stationary.)

So with this dataset we find that ADF indicates non-stationary data, while KPSS shows trend stationary data. Let us return to our stationarity algorithm:

1) **ADF and KPSS conclude that the data is not stationary:** assume not stationary

2) **ADF and KPSS conclude that the series is stationary:** assume strictly stationary

3) **ADF concludes not stationary, KPSS concludes trend stationary:** we have trend stationary data where we need to remove the trend to make the data strictly stationary

4) **ADF concludes stationary, KPSS concludes not trend stationary:** we have difference stationary and can use differencing to make data strictly stationary

In this case, ADF indicates non-stationary data while KPSS indicates trend stationary data. So we can conclude that condition 3 applies. Therefore we can treat the data as trend stationary: the data can be made stationary after we remove the trend.

Because the CO2 concentration dataset is relatively simple and does not involve seasonality, we can render it stable by removing the trend, by means of removing the rolling means from the actual data (a form of smoothing). To do this, we make another column in the CO2_ConcentrationCut data frame called RemoveRollingMean:

```
# smoothing - subtract rolling mean from data
CO2_ConcentrationCut$RemoveRollingMean <-
    CO2_ConcentrationCut$CO2 - CO2_rollmean
```

The subtraction on the right-hand side of the formula creates the differences between the CO2 concentration values and the rolling means. These values are assigned to the new column specified on the left-hand side.

We can use the ADF test to check the stationarity of this new column.

```
adf.test(na.omit
    (CO2_ConcentrationCut$RemoveRollingMean),
    alternative = "stationary", k=12)
```

The first argument in the adf.test() function is the na.omit() function, which deletes the 11 leading NA values from the CO2_ConcentrationCut$RemoveRollingMean column, so we only carry out the ADF test on valid values. From the test output, we can see that the CO2 concentration data with the rolling mean values removed is stationary ($p < .05$):

```
        Augmented Dickey-Fuller Test
data:  na.omit(CO2_ConcentrationCut$RemoveRollingMean)
Dickey-Fuller = -3.5424, Lag order = 12, p-value = 0.04123
alternative hypothesis: stationary
```

If we use the plot() function, we can see that there is no longer an increasing trend. The data is now ready for modelling!

```
plot(CO2_ConcentrationCut$RemoveRollingMean,
     xlab = "Time",
     ylab = "CO2 with rolling mean removed")
```

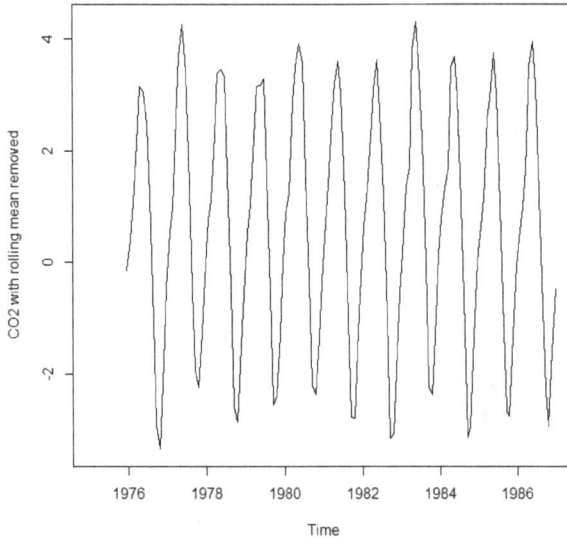

Although we can use smoothing to remove the trend of the CO2 Concentration data because of its simplicity, in real life not all non-stationarity can be so easily rendered stationary. Hence, when we have a more complicated dataset, we need to use differencing and decomposition to make the data stationary. Here, I will perform a brief demonstration of differencing and decomposition using the CO2 concentration dataset.

```
# Perform ADF test
# on 1st Order Differenced CO2 Data
adf.test(diff(CO2_ConcentrationCut$CO2,
     differences=1),
     alternative = "stationary", k=12)
```

Differencing can be carried out by using the diff() function. For the first argument in diff(), we put in the CO2 column of the CO2_ConcentrationCut data frame and for the second argument, we put in differences=1 to indicate that differencing was only performed on one occasion. To test the stationarity of the data after the first differencing, we place the diff() function with its arguments inside the adf.test() function. From the output, we can see that the data is not stationary (the p value is greater than 0.05).

```
        Augmented Dickey-Fuller Test

data:  diff(CO2_ConcentrationCut$CO2, differences = 1)
Dickey-Fuller = -3.2507, Lag order = 12, p-value = 0.08228
alternative hypothesis: stationary
```

When this happens, we try to fix it by increasing the order of the difference, like using differences=2 in the second argument of the diff() function. This time, the data is stationary:

```
# Perform ADF test
# on 2nd Order Differenced CO2 data
adf.test(diff(CO2_ConcentrationCut$CO2,
```

```
differences=2),
alternative = "stationary", k=12)
```

```
        Augmented Dickey-Fuller Test
data:   diff(CO2_ConcentrationCut$CO2, differences = 2)
Dickey-Fuller = -10.309, Lag order = 12, p-value = 0.01
alternative hypothesis: stationary
```

For **decomposition**, we first need to convert the CO2 data from the CO2_ConcentrationCut data frame into a time series format using the ts() function.

```
# Convert CO2 data from Jan 1975 to Dec 1986
# to time series format
TS_CO2Concentration <-
      ts(CO2_ConcentrationCut$CO2,
      start= c(1975, 1), end = c(1986,12),
      frequency = 12)
```

The CO2 column is in the first argument of the function. The start and end dates are the second and third arguments; here, we only use the data from the 1st month of 1975 to the 12th month of 1986 for the conversion. The final argument, the frequency, is stipulated as 12 because of the monthly collection of data. This new time series data is assigned the name TS_CO2Concentration.

```
# Perform decomposition on TS_CO2Concentration
TS_CO2Concentration_decom <-
      decompose(TS_CO2Concentration,
      type = "multiplicative")
```

For decomposition, we use the decompose() function on the time series data. Within the decompose() function, the first argument carries the dataset, TS_CO2Concentration, second being type="multiplicative". The decomposed time series data is named TS_CO2Concentration_decom.

```
# Plot the decomposed data
plot(TS_CO2Concentration_decom)
```

If we use the plot() function on TS_CO2Concentration-_decom, the output graph shows the original data (observed), the trend of the data (trend), the seasonality of the data (seasonal), and the random part of the data after trend and seasonality have been removed.

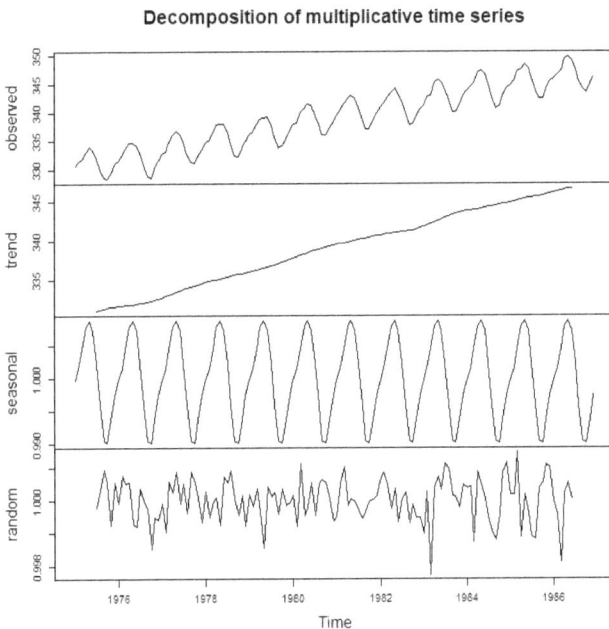

Decomposition of multiplicative time series

ARIMA – Auto-Regressive Integrated Moving Average

Now that we have our stationary time series data, it's time to model it. Before we start with model building, we need to calculate 3 parameters for this model: p, d, and q. **p** is the number of autoregressive (AR) terms, which are the predictors for the outcome variable. **q** is the number of moving average (MA) terms in the model. **d** is the number of differences that were taken when we were trying to make the original data stationary. For the values of p and q, we can find candidate values for them using the autocorrelation function (ACF) for p and partial autocorrelation function (PACF) for q.

For obtaining p and q, we can use the Acf() and Pacf() functions from the forecast package. Because we have found the CO2 data differenced 2 times to be stationary, we will use this to find our ARIMA model, therefore also using it with the ACF and PACF functions to obtain the p and q values.

```
# Creating the ARIMA model
# (using the differenced CO2 TS Data)
# Plotting the ACF and PACF
CO2_ConcentrationCutDiff <-
    diff(CO2_ConcentrationCut$CO2,
    differences=2)
```

First, we use the diff() function to take 2 differences of the CO2 variable in the CO2_ConcentrationCut data frame;

the name CO2_ConcentrationCutDiff has the times series data assigned to it. Then we put the new data into the Acf() and Pacf() functions. Two graphs result from this action and we can obtain the p and q values by looking at the position of the 'lags' in the graph where the values first go from positive to negative.

```
Acf(CO2_ConcentrationCutDiff)
```

Series CO2_ConcentrationCutDiff

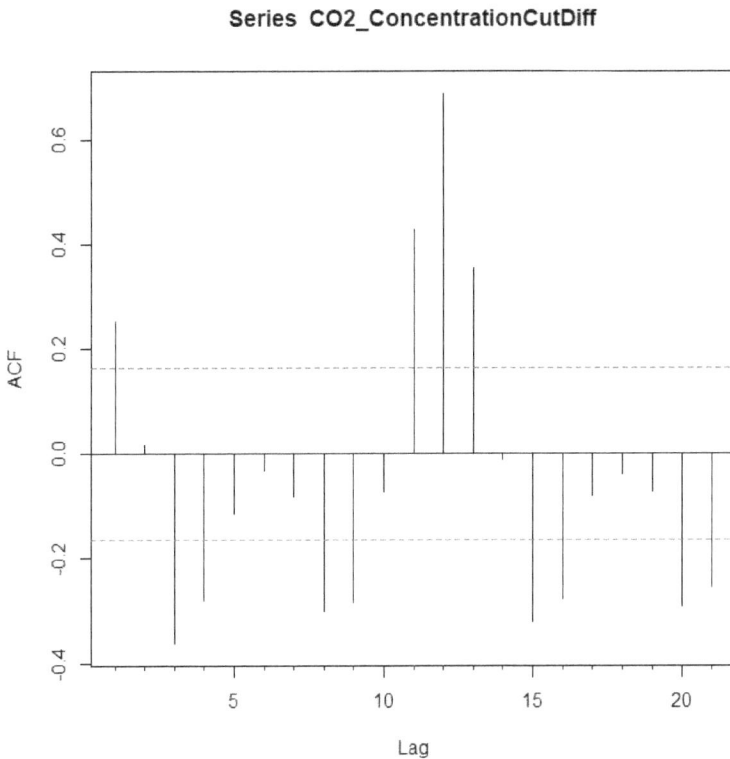

Looking from left to right, on the ACF graph, we see two positive values (the second is tiny, only just discernible above the horizontal line), followed by a negative value. This is at lag 3. The lag number provides us with the value p (not to be confused with significance p). So in this case, p = 3.

```
Pacf(CO2_ConcentrationCutDiff)
```

Series CO2_ConcentrationCutDiff

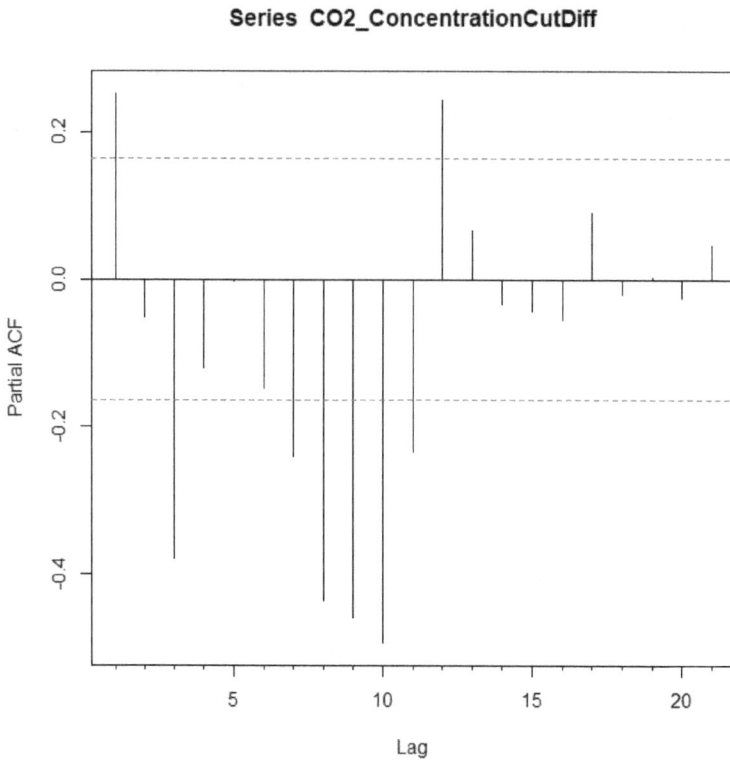

We obtain the q value from the PACF graph. In this instance, the values first move from positive to negative at lag 2, so the q value is 2.

So 3 and 2 are good values for p and q, which we can use to build the ARIMA model. To do this, we need the Arima() function from the forecast package.

```
fit <-
    Arima(CO2_ConcentrationCutDiff, c(3, 2, 2))
summary(fit)
```

The first argument is the differenced time series. The second argument is a vector containing the **p**, **d**, and **q** values (in that order): the **d** value is the degree of differencing (in this case, 2). We assign the name 'fit' to this ARIMA model. We can put the fit model into the summary() function to look at the characteristics of the model and judge how good it is for the dataset by looking at the AIC and BIC values. The lower the AIC and BIC values, the better fit the model is to the data.

```
Series: CO2_ConcentrationCutDiff
ARIMA(3,2,2)

Coefficients:
         ar1      ar2      ar3      ma1     ma2
      0.2584   0.0605  -0.3789  -1.9970   1.000
s.e.  0.0782   0.0819   0.0790   0.0262   0.026

sigma^2 estimated as 0.7944:  log likelihood=-187.54
AIC=387.07   AICc=387.7   BIC=404.72

Training set error measures:
                    ME       RMSE      MAE  MPE MAPE      MASE         ACF1
Training set -0.02963717 0.8690656 0.685912 -Inf  Inf 0.7150727 -0.04031636
```

The 'ar' (autoregression) coefficients are used to illustrate the influence of the previous values on the current value of the variable, while the 'ma' (moving average) coefficients are used to illustrate the noise from the previous values. There are 3 AR coefficients and 2 MA coefficients because we specified 3 for p and 2 for q based on the ACF and PACF tests.

To see the significance of each parameter, we pass the model fit through the coeftest() function (which is from the lmtest package):

```
coeftest(fit)
```

```
z test of coefficients:

     Estimate Std. Error  z value  Pr(>|z|)
ar1  0.258371   0.078172   3.3051 0.0009493 ***
ar2  0.060507   0.081896   0.7388 0.4600075
ar3 -0.378879   0.078970  -4.7978 1.604e-06 ***
ma1 -1.997038   0.026195 -76.2374 < 2.2e-16 ***
ma2  0.999979   0.025974  38.4987 < 2.2e-16 ***
---
signif. codes:  0 '***' 0.001 '**' 0.01 '*' 0.05 '.' 0.1 ' ' 1
```

We can see from the p values of the coefficients that ar1 (at timepoint -1, the timepoint right before the current time point) and ar3 (timepoint -3) are below 0.05, which means they are significant. We should also retain ar2 (timepoint -2), because it doesn't make sense to think of the current timepoint as independent of this value: 'timepoint -2' is dependent on the timepoints before and after it. In the case of ma1 and ma2, we have the noise from the values of timepoint -1 and timepoint -2. The process of detecting stationarity in order to fit the ARIMA model using the p and q values we found using ACF and PACF is known as the **Box-Jenkins method** (Box *et al*, 1994).

Auto ARIMA

Although the Box-Jenkins method will give you a somewhat 'good enough' model, there is a better method that will help you find the best ARIMA model for your time series data. The method is called **Auto ARIMA**, which tests all possible combinations of the p, d, and q values within a certain limit. Auto ARIMA can be conducted in R using the auto.arima() function from the forecast package.

```
auto.fit <-
    auto.arima(CO2_ConcentrationCut$CO2)
summary(auto.fit)
```

The auto.arima() function needs the unmodified time series data CO2_ConcentrationCut$CO2 to find the best ARIMA model; we assign this model the name auto.fit. Using the summary() function, we can take a look at the characteristics of this model.

```
Series: CO2_ConcentrationCut$CO2
ARIMA(2,1,4) with drift

Coefficients:
         ar1      ar2      ma1     ma2      ma3     ma4    drift
      1.6647  -0.9344  -1.2904  0.2418  -0.1616  0.3387   0.1139
s.e.  0.0302   0.0304   0.0800  0.1358   0.1614  0.0925   0.0224

sigma^2 estimated as 0.3306:  log likelihood=-122.92
AIC=261.84   AICc=262.92   BIC=285.55

Training set error measures:
                   ME       RMSE       MAE          MPE      MAPE      MASE
Training set 0.002344842 0.5588205 0.4470885 0.0004660305 0.1318087 0.4063149
                   ACF1
Training set -0.02837766
```

The AIC and BIC of this model are both lower than the model we built using the Box-Jenkins method, which means that it is a better model for the data. This is also a great method for comparing which level of differencing was more effective on the model. From the second line of the output, you can see that this model has different p, d, and q (2, 1, and 4 instead of 3, 2, and 2) values than the Box-Jenkins method model and it is with 'drift' (used to accommodate underlying increasing of data not eliminated after differencing).

We need to be cautious in model interpretation situations where the Box-Jenkins model is the only option (although this will probably never happen because as long as the Box-Jenkins model can be built, then so can the auto ARIMA model). As for the coefficients of the ar and ma coefficients, since time series models are mainly for the purpose of modeling the variable value at the current timepoint relative to the previous (thereby using it to forecast future values), the coefficients for ar and ma coefficients are not frequently used for model interpretation.

```
coeftest(auto.fit)
```

```
z test of coefficients:

        Estimate Std. Error  z value  Pr(>|z|)
ar1     1.664650   0.030153  55.2064  < 2.2e-16 ***
ar2    -0.934373   0.030371 -30.7655  < 2.2e-16 ***
ma1    -1.290434   0.080021 -16.1262  < 2.2e-16 ***
ma2     0.241787   0.135815   1.7803  0.0750318 .
ma3    -0.161630   0.161355  -1.0017  0.3164850
ma4     0.338665   0.092508   3.6609  0.0002513 ***
drift   0.113856   0.022351   5.0939  3.507e-07 ***
---
Signif. codes:  0 '***' 0.001 '**' 0.01 '*' 0.05 '.' 0.1 ' ' 1
```

Now that we have built our ARIMA model, we can use it to make a prediction. Since the auto ARIMA model is the better one, we will use it for this purpose. To do this, we use the forecast() function from the forecast package.

```
pred <- forecast(auto.fit, h=24)
plot(pred, xlab = "Time Point", ylab = "CO2",
     fcol = gray.colors(1))
```

For the first argument in the forecast() function, we use the name of our model, auto.fit, and for the second argument, we enter h=24 to predict the concentration of CO2 in the air over the next 24 months (comprising the monthly concentration of CO2 in the air for 1987 and 1988). We assign the prediction the name 'pred'. Then we plot our prediction using the plot() function where pred is the first argument, then labelling the x and y axes as "Time Point" and "CO2" respectively. The last argument is the color of the forecast ('fcol'); gray.colors(1) ensures that the forecast is a gray color.

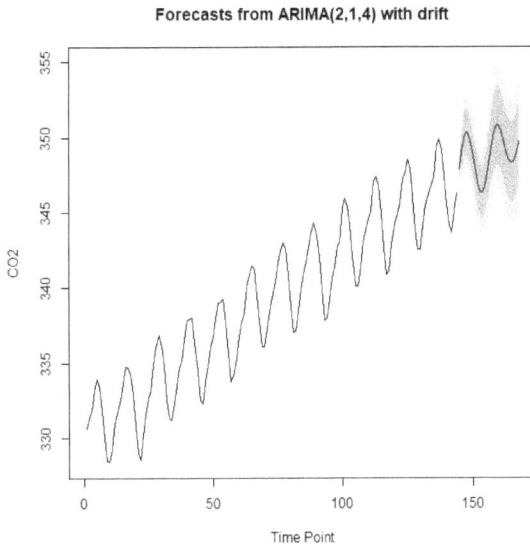

Forecasts from ARIMA(2,1,4) with drift

```
pred
```

```
    Point Forecast      Lo 80      Hi 80      Lo 95      Hi 95
145        347.8938   347.1569   348.6307   346.7668   349.0208
146        349.1545   347.9021   350.4070   347.2391   351.0700
147        350.1218   348.5429   351.7007   347.7071   352.5365
148        350.3745   348.7186   352.0304   347.8421   352.9070
149        349.9221   348.2660   351.5782   347.3893   352.4549
150        348.9636   347.2757   350.6516   346.3821   351.5452
151        347.8215   346.0681   349.5748   345.1400   350.5029
152        346.8465   345.0503   348.6426   344.0995   349.5934
153        346.3214   344.5218   348.1210   343.5692   349.0736
154        346.3890   344.5687   348.2093   343.6051   349.1729
155        347.0229   345.0886   348.9571   344.0647   349.9811
156        348.0456   345.9003   350.1909   344.7647   351.3266
157        349.1865   346.8077   351.5654   345.5484   352.8247
158        350.1609   347.6035   352.7182   346.2498   354.0720
159        350.7475   348.0969   353.3980   346.6937   354.8012
160        350.8442   348.1669   353.5216   346.7496   354.9388
161        350.4880   347.8097   353.1662   346.3919   354.5840
162        349.8352   347.1533   352.5170   345.7336   353.9367
163        349.1121   346.4225   351.8016   344.9988   353.2254
164        348.5491   345.8571   351.2410   344.4321   352.6661
165        348.3182   345.6246   351.0118   344.1987   352.4377
166        348.4907   345.7733   351.2081   344.3348   352.6465
167        349.0242   346.2407   351.8077   344.7672   353.2812
168        349.7818   346.8955   352.6681   345.3676   354.1961
```

The first column shows exact predicted values (the first column) as well as 80% and 95% confidence intervals, demonstrating the likely limits of predictive values. The hi and lo 80s are represented as the darker gray area surrounding the prediction line while the hi and lo 90s are represented by the lighter gray area as well as the darker gray area.

Now that we have discussed time series data, how to make them stationary, how to build ARIMA models, and how to use your models to make predictions, it is time for you to find some simple time series data (without seasonality) of your own and apply your new skills. However, if you would like more of a challenge, the next section involves analyzing time series data which involves seasonality. The techniques for analyzing time series data with and without seasonality are quite similar. Therefore, it is quite easy to indulge in your adventurous side and learn about seasonal ARIMA (SARIMA), applying all of your skills to datasets of your choice.

Seasonal ARIMA (SARIMA)

Now that we have gone through how ARIMA works for non-seasonal time series data, what about seasonal time series data? Seasonality is a characteristic that adds complexity to time series data and it occurs frequently in real-life time series data. Therefore, learning how to handle it is important. The process is similar to a time series without seasonality. The dataset we are going to use for the example is the Monthly Auto Sales data from Jan 1967 to Jan 2021 in the United States.

```
USMonthlyAutoSales <-
    read.csv("USMonthlyAutoSales.csv")
USMonthlyAutoSales$SALES <-
    ts(USMonthlyAutoSales$SALES,
    frequency = 12)
```

We start by reading the dataset into the R environment using the read.csv() function and call it USMonthlyAutoSales. The dataset is composed of 2 columns: DATE and SALES. We first convert the SALES column to time series data using the ts() function where the first argument is the SALES column and the second argument is frequency, set at 12 because it is monthly sales data. On the left-hand side, we assign the converted time series data to the SALES column of the USMonthlyAutoSales dataset to replace the original SALES column.

```
plot(USMonthlyAutoSales$SALES, type = "l",
    ylab="US Monthly Auto Sales",
    xlab = "Time Points")
```

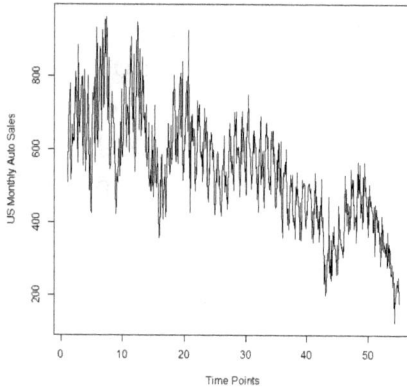

Plotting the SALES column, we can see that it is very difficult to tell visually the stationarity of the dataset. Therefore, we need to use the ADF and KPSS tests to determine the stationarity of the data.

```
adf.test(USMonthlyAutoSales$SALES)
```

```
        Augmented Dickey-Fuller Test

data:  USMonthlyAutoSales$SALES
Dickey-Fuller = -4.4896, Lag order = 8, p-value = 0.01
alternative hypothesis: stationary
```

(A high p value for the ADF test indicates non-stationary data; $p < 0.05$ indicates stationary data.)

```
kpss.test(USMonthlyAutoSales$SALES,
    null = "Trend")

        KPSS Test for Trend Stationarity

data:  USMonthlyAutoSales$SALES
KPSS Trend = 0.2187, Truncation lag parameter = 6, p-value = 0.01
```

(A high p value for the KPSS test indicates trend stationary; $p < .05$ indicates non-stationary.)

From the outputs, the ADF test output shows that the data is stationary and the KPSS test output shows that the data is not trend stationary. Let's look at the key:

1) **ADF and KPSS conclude that the data is not stationary:** assume not stationary

2) **ADF and KPSS conclude that the series is stationary:** assume strictly stationary

3) **ADF concludes not stationary, KPSS concludes trend stationary:** we have trend stationary data where we need to remove the trend to make the data strictly stationary

4) **ADF concludes stationary, KPSS concludes not trend stationary:** we have difference stationary and can use differencing to make data strictly stationary

Condition 4 applies in this case, so the data can become strict stationary if we use differencing.

```
sarima1 <- arima(USMonthlyAutoSales$SALES,
  order = c(0, 1, 0),
  seasonal = list(order=c(0, 0, 0), period=12))
```

Following the results of the ADF and KPSS tests, we begin by applying 1 order of differencing to the data and build a SARIMA model using the differenced data. We use the arima() built-in function: the first argument is the SALES column of the dataset. The second argument, order, controls the values of the p, d, q parameters of the model (the values are contained in a vector). In this case, the d value is 1, so we have c(0, 1, 0). The third argument, seasonal, controls the parameters of the seasonality part of the SARIMA model. These three parameters are represented as P, D, and Q, capital versions of the letters used to represent the ARIMA parameters (Hyndman and Athanasopoulos, 2018), and they are also entered using a vector.

Because we are starting with only 1 order of differencing for now, the vector we enter for 'seasonal' is c(0, 0, 0). The 'seasonal' argument also takes the number of periods in the time series data into consideration; as the dataset contains the number of automobiles sold monthly, the period is 12.

The list() function allows us to enter the seasonal parameters and the number of periods together. We assign this SARIMA model the name sarima1.

```
plot(residuals(sarima1))
```

To see how well this SARIMA model fits the USMonthly-AutoSales data, we can take a look at the residuals – the parts of the dataset not accounted for by the sarima1 model – using the residuals() function.

As well as plotting the residuals, we can also look at their autocorrelation using the Acf() and Pacf() functions.

```
Acf(residuals(sarima1))
```

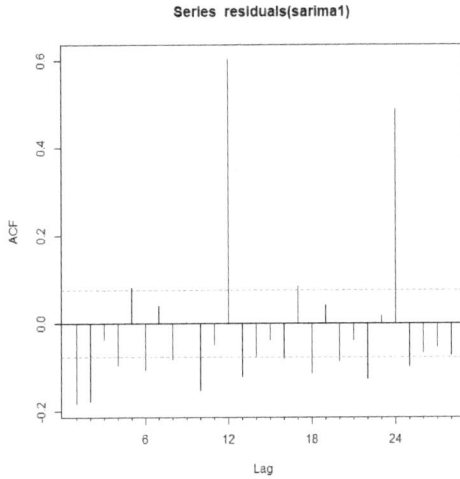

Series residuals(sarima1)

```
Pacf(residuals(sarima1))
```

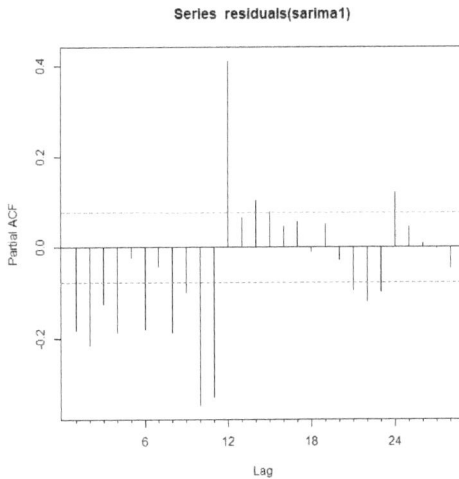

Series residuals(sarima1)

From the ACF and PACF outputs, we can see that the autocorrelation is higher at the timepoints 12 and 24, which implies that there is autocorrelation at each period. Therefore, we can try adding an order of differencing for the seasons, naming this new model sarima2:

```
sarima2 <- arima(USMonthlyAutoSales$SALES,
   order = c(0, 1, 0),
   seasonal = list(order=c(0, 1, 0), period=12))
plot(residuals(sarima2))

Acf(residuals(sarima2))
```

Series residuals(sarima2)

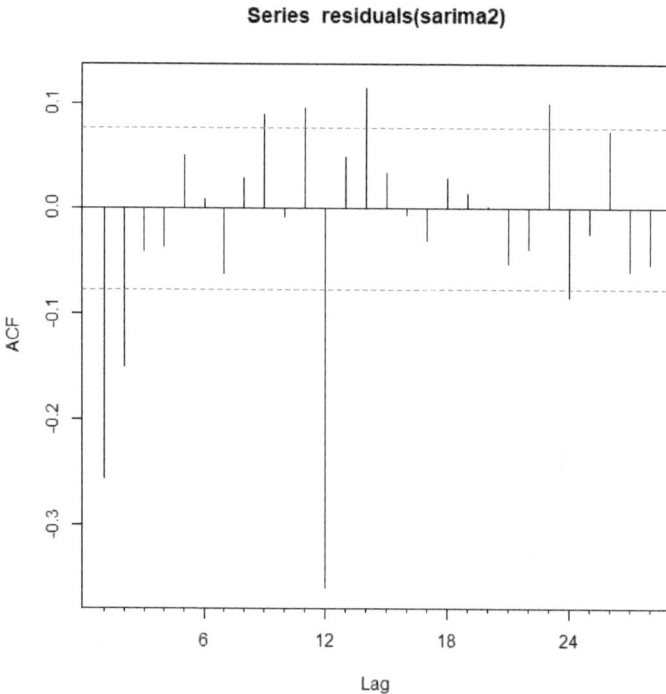

```
Pacf(residuals(sarima2))
```

Series residuals(sarima2)

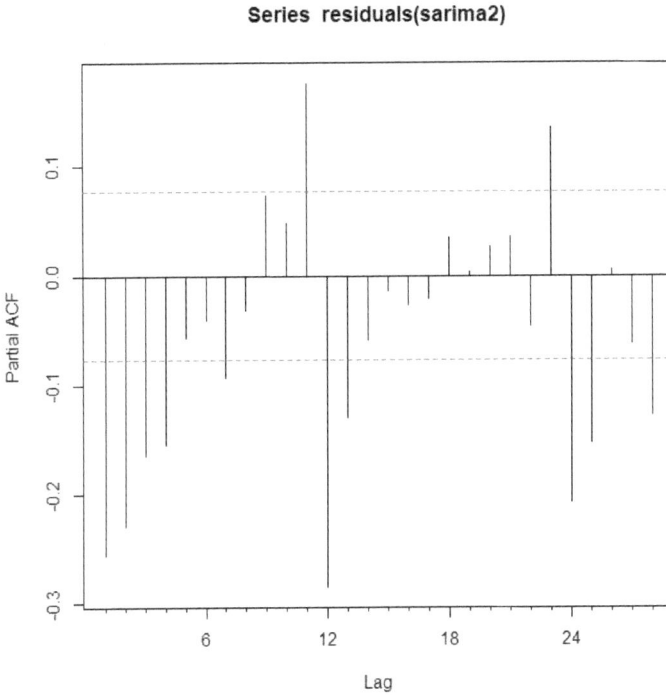

From the ACF of the residuals of sarima2, we can see that the ACF is significantly negative at lag 2, and at lag 3 the autocorrelation is insignificant. Hence, we can try to use 2 as the p value and name the new model sarima3.

```
sarima3 <- arima(USMonthlyAutoSales$SALES,
   order = c(2, 1, 0),
   seasonal = list(order=c(0, 1, 0), period=12))

Acf(residuals(sarima3))
```

Series residuals(sarima3)

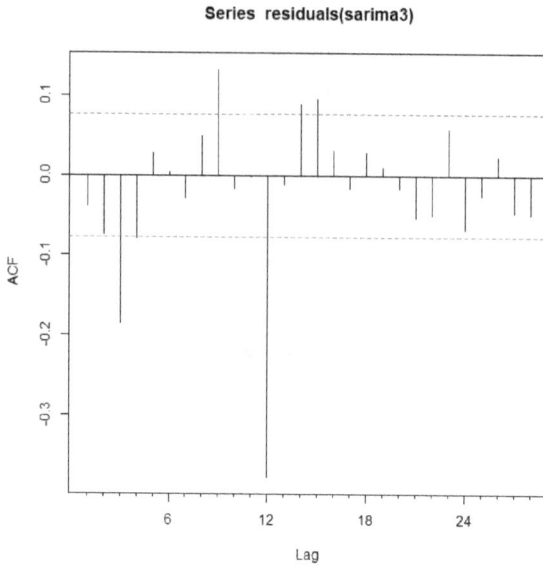

Pacf(residuals(sarima3))

Series residuals(sarima3)

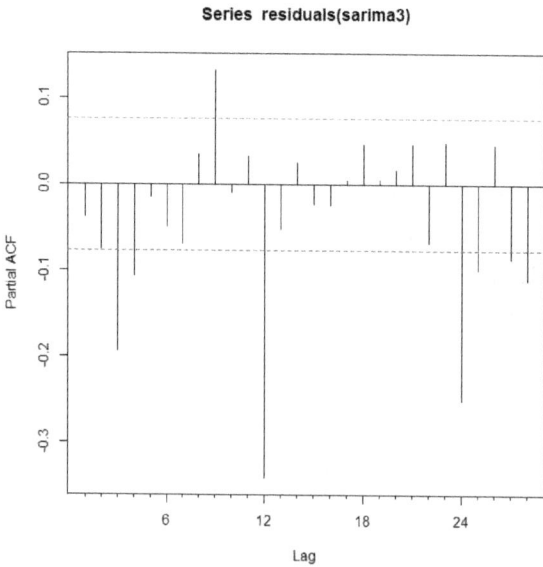

From the ACF and PACF of the residuals of sarima3, we can see that the autocorrelations are insignificant at most of the lags. However, the autocorrelations at 12 and 24 are still large and significant. Therefore, we should modify the seasonal parameters P and Q to control this seasonal autocorrelation. First we increase the seasonal parameter P by 1 and name this new model sarima4.

```
sarima4 <- arima(USMonthlyAutoSales$SALES,
   order = c(2, 1, 0),
   seasonal = list(order=c(1, 1, 0), period=12))

Acf(residuals(sarima4))
```

Series residuals(sarima4)

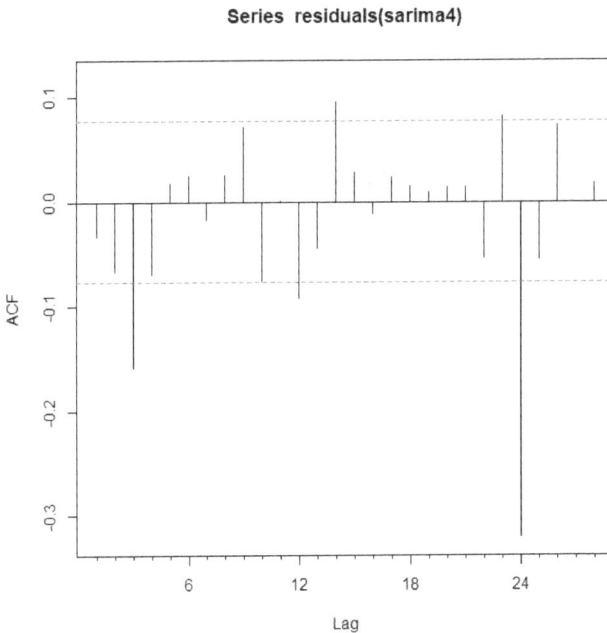

```
Pacf(residuals(sarima4))
```

Series residuals(sarima4)

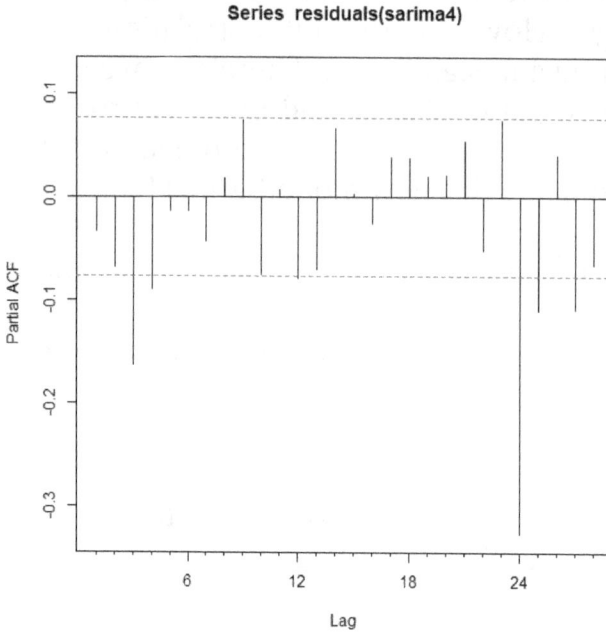

From the ACF and PACF graphs of sarima4, we can see that the autocorrelation at lag 12 is now less than in sarima3 and the autocorrelations at other lags are still mostly insignificant. However, the autocorrelation at lag 24 is highly significant. Let's increase seasonal parameter Q by 1 and call this new model sarima5.

```
sarima5 <- arima(USMonthlyAutoSales$SALES,
  order = c(2, 1, 0),
  seasonal = list(order=c(1, 1, 1), period=12))

Acf(residuals(sarima5))
```

Series residuals(sarima5)

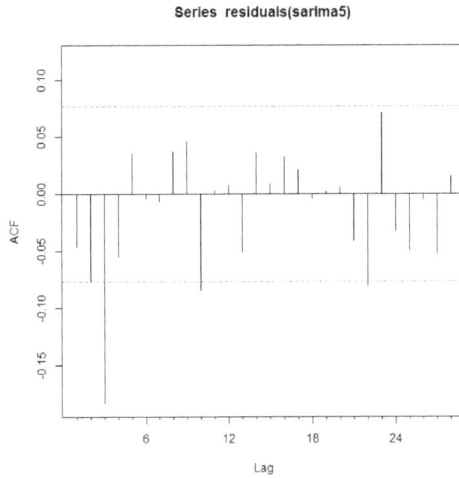

Pacf(residuals(sarima5))

Series residuals(sarima5)

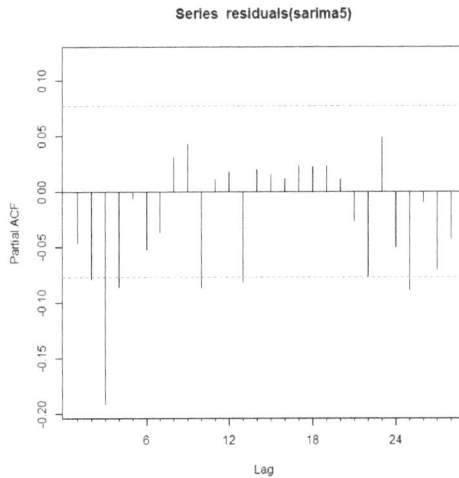

From the ACF and PACF graphs of sarima5, we can see that there is still a significant autocorrelation at lag 3, but other than that, most of the lags no longer have significant autocorrelations. We can now compare sarima5 with the Auto ARIMA model. First, we can run the model name sarima5 alone to get the coefficients and AIC of the model. Then, we can check the significance of the coefficients using the coeftest() function of the lmtest package.

```
sarima5
```

```
call:
arima(x = USMonthlyAutoSales$SALES, order = c(2, 1, 0), seasonal = list(order = c(1,
    1, 1), period = 12))

Coefficients:
          ar1      ar2     sar1     sma1
      -0.3257  -0.2612   0.1583  -0.8418
s.e.   0.0386   0.0385   0.0495   0.0280

sigma^2 estimated as 2901:  log likelihood = -3443.78,  aic = 6897.55
```

```
coeftest(sarima5)
```

```
z test of coefficients:

        Estimate Std. Error  z value  Pr(>|z|)
ar1    -0.325694   0.038624  -8.4325 < 2.2e-16 ***
ar2    -0.261162   0.038485  -6.7861 1.152e-11 ***
sar1    0.158287   0.049541   3.1951  0.001398 **
sma1   -0.841792   0.028043 -30.0178 < 2.2e-16 ***
---
signif. codes:  0 '***' 0.001 '**' 0.01 '*' 0.05 '.' 0.1 ' ' 1
```

Next, we can build the Auto ARIMA model using the auto.arima() function:

```
auto.sarima <-
    auto.arima(USMonthlyAutoSales$SALES,
    seasonal = T)
```

The first argument contains the SALES column of the US-MonthlyAutoSales dataset and the second argument in-

---

cludes seasonality as part of the model. We call this model auto.sarima and run the name alone to get the coefficients and AIC. We can also use the Acf() and Pacf() functions on the residuals of auto.sarima to get the ACF and PACF graphs for comparison purposes as well as use the coeftest() function on auto.sarima to look at the significance of the parameters.

```
auto.sarima
```

```
Series: USMonthlyAutoSales$SALES
ARIMA(1,0,2)(0,1,2)[12] with drift

Coefficients:
          ar1      ma1      ma2     sma1     sma2    drift
       0.9494  -0.3760  -0.1982  -0.6619  -0.1597  -0.6085
s.e.   0.0168   0.0428   0.0401   0.0395   0.0386   0.2786

sigma^2 estimated as 2701:  log likelihood=-3423.64
AIC=6861.28    AICc=6861.46    BIC=6892.48
```

```
Acf(residuals(auto.sarima))
```

Series residuals(auto.sarima)

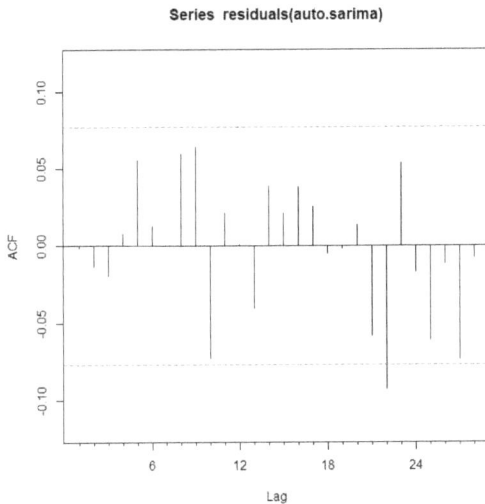

```
Pacf(residuals(auto.sarima))
```

Series residuals(auto.sarima)

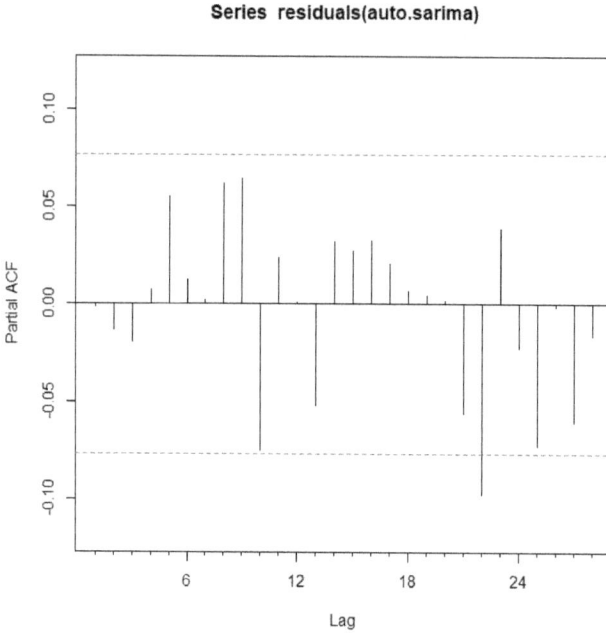

```
coeftest(auto.sarima)
```

```
z test of coefficients:

        Estimate Std. Error   z value  Pr(>|z|)
ar1     0.949350   0.016789  56.5472 < 2.2e-16 ***
ma1    -0.375975   0.042763  -8.7920 < 2.2e-16 ***
ma2    -0.198223   0.040083  -4.9453 7.603e-07 ***
sma1   -0.661854   0.039545 -16.7368 < 2.2e-16 ***
sma2   -0.159693   0.038638  -4.1330 3.580e-05 ***
drift  -0.608498   0.278580  -2.1843   0.02894 *
---
signif. codes:  0 '***' 0.001 '**' 0.01 '*' 0.05 '.' 0.1 ' ' 1
```

From the coefficient outputs of sarima5 and auto.sarima, we can see that all coefficients are significant in both models (with sarima5 having 4 parameters while auto.sarima has 6). Considering the ACF and PACF graphs of each model, auto.sarima had overall less lags with significant autocorrelations. For AIC, auto.sarima had a lower AIC of 6861.46 then sarima5, with an AIC of 6897.55. When choosing a better time series model, we want a model with significant parameters, less parameters (by rule of parsimony), less autocorrelation in the residuals, and lower AIC. In this case, auto.sarima is the better model.

Some of you may ask "April, since in the previous two examples, the models developed by the auto.arima() function are both superior than the models developed using the Box-Jenkins method, can we forego building our own models and directly use auto.arima() to build time series models?" The answer is yes and no. The auto.arima() function is used to build a reference model of time series data where it returns the model with the lowest AIC without considering the other factors. It is also possible at times that it will return a model with more parameters and higher AIC than a model built using the Box-Jenkins method. If there is sufficient time and resources, it is best to build our own models and compare them with an auto.arima() model.

Further reading

Now that you have learned about time series data and how to build models, you may think that time series model building may be more arbitrary than other methods. As George Box (as in Box-Jenkins' model) once said, "All models are wrong, but some are useful." In order to build better (and more useful) models for any time series data you may encounter in the future, it is best to educate yourselves more deeply in the subject. Two great reference books are *Time Series Analysis and Its Applications* by Shumway and Stoffer (2017) and *Forecasting: Principles and Practice* by Hyndman and Athanasopoulos (2018). Studying time series analysis more deeply and practicing on some times series data of your choice, you will be able to build your own wrong but useful time series models.

Chapter 3 – Survival Analysis

Sections include

Non-parametric methods
    Life tables
    Kaplan-Meier (aka Product Limit)

Semi-parametric method
    Cox proportional hazards model

Parametric methods
    Weibull distribution
    Exponential distribution

Introduction

Now that we have gone through the complexity of structural equation modelling and time series analysis, you may be very excited to see the title of this chapter because 'survival analysis' sounds like something that will help you survive a difficult data analysis situation. Although survival analysis is a very useful branch of statistics, it is not the Holy Grail of statistics. **Survival analysis** (aka **time-to-event analysis**) is a branch of statistics used to analyse time-to-event data, where the data in question describes how much time it took to reach a certain event (Kleinbaum and Klein, 2006).

Examples include the survival time of a patient diagnosed with cancer from time of diagnosis to death; the time it takes for a battery to run out of juice since its creation, or the time it takes for a person suffering from mathephobia to become a reasonably good data analyst or statistician after they got their hands on this book! Note that **events** can be negative (such as deaths) or positive (promotions, graduations etc).

Survival analysis is widely used in public health, clinical studies, and in industry. To best understand the nature of survival analysis, let's first look at a simple example. Let's say that we recruited a group of 6 employees and followed them for 10 years after their first day at the company to see how long it takes for each employee to change jobs. The chart below represents the tenure time for each employee who joined the study. One thing we should note is that, although in the graph it seems as though all 6 employees joined the company at the same time, this may not necessarily be the case. The employees may join the study at any time and we can attempt to follow them for 10 years starting from their time of joining the study. But in a graph, we can make it *seem* as though they have joined at the same time because the time they joined the study is irrelevant as long as we have attempted to track them for the entire 10-year study time.

Number of Years New Employees Stay At Their New Jobs Before Job Change

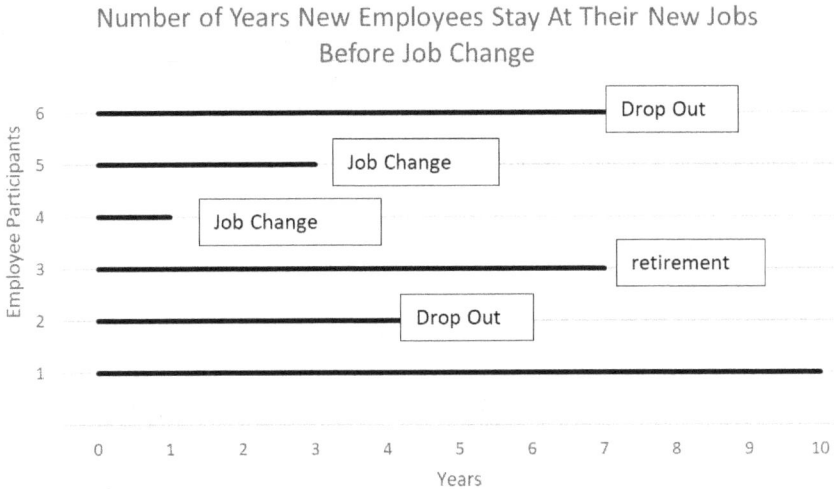

From the chart, we can see that of the six new employees, employees 4 and 5 changed their jobs; in other words, they experienced an **event** within the 10 year period, at years 1 and 3 respectively. In survival analysis terminology, their **times-to-event** (or **survival times**) are 1 and 3 years respectively. Employee 1 did not change job during the entire 10 year period. Employees 2 and 6 dropped out of the study and employee 3 retired from the workforce before their 10 year follow-up was over. These employees did not change jobs during their shortened participation time in the study. For these subjects, who are 'lost to follow-up', the data they contributed are called **censored data**. Censored data can still be used in a survival study because they contribute information, that the subject did not experience an event up to a certain amount of time. However, this is under the condition that the censored data are non-informative censoring, where we assume that the censored data would have the same distribution (or overall pattern) as the non-censored if they had participated until the end of the 10 year period. Another assumption we make in order to use the censored data in the study is that the reason for the participant leaving the study does not influence the chance of the subjects developing the event in question. For the rest of this chapter, all the censored data presented are non-informative censoring.

Three methodological branches have been developed to analyze survival (or time-to-event) data: 1) Non-parametric, 2) Semi-parametric, and 3) Parametric (Kleinbaum and Klein, 2006). If any of the terms mentioned above make you think "April, I am confused by these terms and by distribution in general. What if I choose the wrong method for analyzing my own data?" - Don't fret, I have put a mind map below to portray the most useful and widely used survival analysis methods and to which of the three branches they belong. The branch of methods we choose to analyze survival data depends on the type of information we hope to obtain from the data and how well we can safely assume the underlying distribution of the survival data.

Non-parametric methods are the methods we use if we do not have knowledge of the underlying distributions of the survival data. The methods under this branch of survival analysis include life table and Kaplan-Meier methods, which are used to compare the survival likelihood of groups of individuals throughout the study time period.

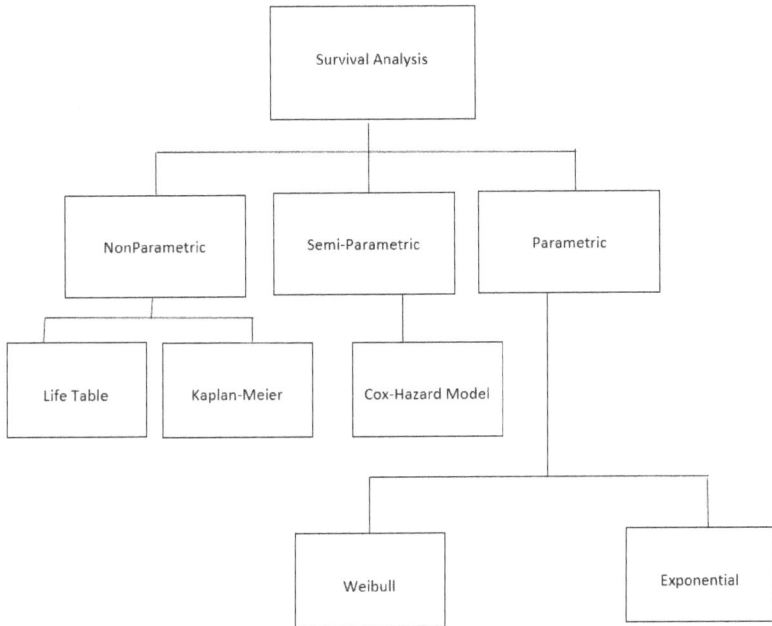

Semi-parametric methods involve knowing the distribution of part of the data but not all. The Cox-hazard model, which is used to determine whether certain variables (conditions) have a significant influence on the survival time of the subjects, is a widely used semi-parametric method for survival analysis.

Parametric methods are methods we can implement when we know the underlying distribution of the survival data of a study. Methods that fall under this branch include Weibull and exponential methods (Hosmer and Lemeshow, 1999), which are covered in this chapter.

Non-parametric survival methods: Life Table and Kaplan-Meier

Life Table method

The **life table** method (aka **actuarial table** method) is an elementary method for analyzing survival data when we do not know the distribution of the survival data. It is also a great place to start when we want a deeper understanding of survival data. It allows us to keep track of the percentage of subjects that are still alive at each time point during the entire study time period (in other words, the survival rate). Like its name, the method involves constructing a life table which serves as the dataset of the analysis. The method assumes that the events of interest occur at the end of the interval and the censored events occur evenly throughout the time interval (Collett, 2003). The life table is constructed by dividing the period of time spent tracking the survival of the subjects into equal intervals and recording the values of each of the following variables at each interval:

N_t – the number of subjects who did not experience events at the beginning of time interval t

D_t - the number of subjects who experienced an event during time interval t

C_t – the number of subjects who were censored during time interval t

NtAvg – the average number of subjects who did not have events during a time interval. Under the assumption that events and censoring occur at the end of the time intervals and evenly throughout the respective intervals, this is calculated by $NtAvg = Nt - (Ct/2)$

PtEvent – the proportion of subjects who do not experience an event during time interval t, $PtEvent = Dt/NtAvg$

PtNoEvent – the proportion of subjects who experience events during time interval t, $PtNoEvent = 1 - PtEvent$

St – the proportion of subjects who survive past time interval t, calculated as $St+1 = Pt+1NoEvent \times St$ where $S_0 = 1$. (S_0 is when t_0, when the survival timeline hasn't started yet. At that time, the survival is 100%, which is 1.) This is simpler than it looks, and will be shown in practice in the walk-through shortly after you see the 'head' of the dataset.

To illustrate the Life Table method, let's look at an example! The table below is the first 6 lines of a survival dataset studying the amount of time it takes for employees of a large company to receive a promotion, where each candidate was followed for 14 years and each year is a time interval. To obtain this, we first read the dataset into the R environment using the read.csv() function on the directory of the dataset and assign the dataset the name PromotionsDataset. Then we apply the head() function to PromotionsDataset to obtain the first six lines of the dataset.

```
PromotionsDataset <- read.csv("Promotions.csv")
head(PromotionsDataset)
```

	Intervals	NumberOfCandidates	Promotions	NumberCensored	City	AvgNumberAtRisk
1	1	4382	100	292	1	4236.0
2	2	3990	750	421	1	3779.5
3	3	2819	965	30	1	2804.0
4	4	1824	263	48	1	1800.0
5	5	1513	279	32	1	1497.0
6	6	1202	88	12	1	1196.0

	ProportionEvent	ProportionNoEvent	SurvivalProb
1	0.024	0.976	0.976
2	0.198	0.802	0.783
3	0.344	0.656	0.513
4	0.146	0.854	0.438
5	0.186	0.814	0.357
6	0.074	0.926	0.330

This dataset contains 9 columns. The column *Intervals* identifies the time intervals. *City* is which of the two company locations is attended by the employees (where 1 = Bigcity and 2 = Smallercity). The other variables are NumberOfCandidates (N_t) at the start of the interval; Promotions (D_t), which are the events; NumberCensored (C_t); AvgNumberAtRisk (N_tAvg); ProportionEvent (P_tEvent); ProportionNoEvent($P_tNoEvent$); and SurvivalProb(S_t). While the dataset in this book is provided in a convenient format, it's important to note that in real life, not all datasets will be ready-made. In some cases, it may be necessary to modify the data into this format before beginning your analyses.

So, let's have a walk-through of how the dataset is created. In the first interval, in the first city, there are 4382 candidates (subjects, Nt). In that interval, 100 are promoted (experience the event, Dt) and 292 are censored (lost to follow-up, Ct). If we take the 4382 and subtract both the promoted employees and those censored (100 + 292 = 392), we get the number of remaining candidates (Nt) in the second row, 3990. This repeats until the next grouping in City when the calculation starts again.

To get the average number at risk (NtAvg), we halve the number censored (Ct/2; 292/2 = 146), then subtract that from the candidates at the start: (subjects, Ct): 4382 − (292/2) = 4236.

Then we calculate ProportionEvent (PtEvent). This is calculated by Promotions (those experiencing the event, Dt) divided by the Average (NtAvg):   100/4236 = 0.024

ProportionNoEvent (PtNoEvent) is calculated by 1 - ProportionEvent (PtEvent), which is 1 −0.024 = 0.976

The last variable, SurvivalProb (St) needs to be handled with care:

```
ProportionNoEvent  SurvivalProb
        0.976          0.976
        0.802          0.783
        0.656          0.513
        0.854          0.438
        0.814          0.357
        0.926          0.330
```

As survival is initially 100%, the first instance of SurvivalProbability is exactly the same as the first instance of ProportionNoEvent (PtNoEvent); both are 0.976. What we next do is to multiply the second case of ProportionNoEvent with the first case of SurvivalProbability: 0.802 * 0.976 = 0.783   The number on the right becomes the second case of SurvivalProbability. And the same happens each time, in a game of catch-up. So the third case of ProportionNoEvent, 0.656, is multiplied by the second case of SurvivalProbability, 0.783, to become, ta-dah! the third case of SurvivalProbability, 0.513.

Hot off the press! The file *Promotions_workings.xls* will make your life much easier. [*]

We use the life table within PromotionsDataset to obtain a time-to-promotion-completion plot (known generically as a **survival plot**). In order to examine the differences in patterns between employees at Bigcity and those at Smallercity during the 14 year follow-up period, we can separate the dataset into two datasets: we take the data from all of the columns from the dataset, but according to whether the *City* variable is equal to 1 or 2, respectively. We can use the city names to name each of the separated datasets:

```
file = PromotionsDataset
Bigcity <- file[file$City==1, ]
Smallercity <- file[file$City==2, ]

plot(Bigcity$Intervals, Bigcity$SurvivalProb,
   type = "s", col = "black",
   xlab = "Intervals (Years)",
   ylab = "Promotion (survival) probability of
   Bigcity and SmallerCity employees")

lines(Smallercity$Intervals,
   Smallercity$SurvivalProb,
   type = "s", col = "black", lwd = 4)
legend("bottomleft",
   c("Bigcity", "Smallercity"), lwd=c(1,4))
```

---

[*]The formulae under the figures build the model from just the original number of employees, the events (Promotions), the censored data, and any grouping (City). Once you get started, the system gets automated and you can adapt this to your own studies.

After that, we create the survival (time-to-event) plot for each city by plotting the survival probability columns of each of the Bigcity and Smallercity datasets using the *plot*() and *lines*() functions. We can start by plotting the survival plot for the Bigcity employees. The first two arguments of the plot() function are the Intervals and SurvivalProbab columns of the Bigcity dataset because they contain the x and y values of the plot respectively. For type (type of graph), we choose type = "s" to create a step plot in order for us to see the sudden drops in the probability of gaining promotions as time goes on. For col (color), I have chosen to make the graph black. And for the xlab and ylab (the x and y axes of the graph), we create appropriate labels.

To add a survival plot of the Smallercity employees to that of the Bigcity employees, we use the lines() function. As in the plot() function, the first two arguments contain the Intervals and SurvivalProbability columns, but from the Smallercity dataset, with the same graph type and color. The additional argument, lwd = 4, makes the plot of the Smallercity employees four times as thick as for the Bigcity employees. The *legend*() function allows us to match the cities' names with the respective line widths. And for those who like to produce things funky, colours could have been added to the mix, by replacing the *legend*() expression with:

```
legend("bottomleft",c("Bigcity","Smallercity"),
    lwd=c(1,3), col=c("red", "blue") )
```

(The resulting chart is not reproduced here, but may prove useful with multiple groups, with or without line widths.)

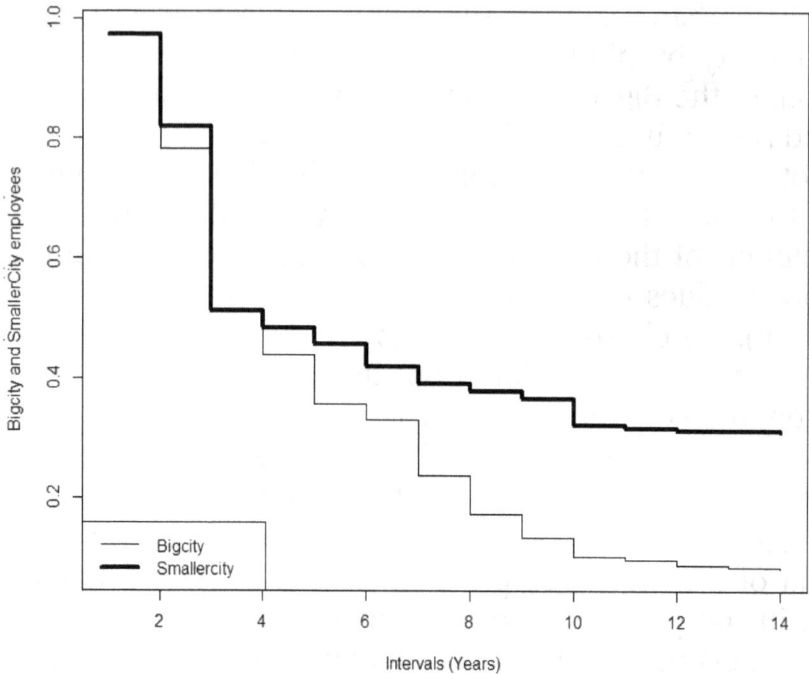

From the survival (time-to-promotion) plots of the employees at the Bigcity and Smallercity locations, we can derive answers to certain questions such as "What proportion of employees in each city need more than 6 years before they are promoted?" Or "What is the median time of promotions in each city location?" For the first question, the answer from the plot is approximately 0.33 for Bigcity students and approximately 0.42 for Smallercity employees because if we go to year 6 and look up until we hit the horizontal lines for Bigcity and Smallercity, the lines are at approximately 0.33 and 0.42 on the y-axis (if you go into the file, you can look up the survival probabilities (Survivalprob)

on interval 6 of the respective groups). For the second question, we can go to the median of the promotion probability in the y-axis (0.5) and read right to find the year where a straight line from 0.5 hits the survival plots of the cities. The median time for employees to get promoted is about 3 years in both occasions; using the file for accuracy, look up the nearest figure to 0.5 in each group, and then look at the corresponding intervals.

Kaplan-Meier method (aka Product Limit method)

Although the life table is a well-designed approach for examining the time-to-event patterns of subjects, there is one major caveat: because the time intervals were decided before the calculation of the survival probability, the survival probability (thereby the survival plots) can change if the time interval changes. In order to address this limitation, the Kaplan-Meier (or product limit method) was created. The Kaplan-Meier method can bypass the limitation of having an interval-dependent survival probability by setting a time interval only when there is an event or censor, and can calculate the survival probability every time there is an event (Kaplan and Meier, 1958).

To illustrate the qualities of a time-to-event dataset designed for the Kaplan-Meier method, we can look at a dataset obtained from 40 couples with difficult marriages (who chose to receive or not receive marriage counselling) and how many of them chose to divorce during a 25 year follow-up period.

```
DivorceCounsel <-
      read.csv("Divorce_Counsel.csv")
DivorceNoCounsel <-
      read.csv("Divorce_NoCounsel.csv")
head(DivorceCounsel)
```

```
   Years Counsel NumAtRisk div censor SurvivalProbability
1 0.0000     NA        20  NA     NA           1.0000000
2 1.4180      1        20   0      1           1.0000000
3 2.8340      1        19   1     NA           0.9473684
4 3.2525      1        18   1     NA           0.8947368
5 3.8690      1        17   0      1           0.8947368
6 4.0850      1        16   0      1           0.8947368
```

```
head(DivorceNoCounsel)
```

```
   Years Counsel NumAtRisk div censor SurvivalProbability
1 0.0000     NA        20  NA     NA                1.00
2 0.4985      0        20   1     NA                0.95
3 0.7500      0        19   1     NA                0.90
4 3.1295      0        18   1     NA                0.85
5 3.7480      0        17   1     NA                0.80
6 5.8480      0        16   0      1                0.80
```

The divorce is the event. Survival Probability is worked out by 1 - (event / numberAtRisk) = 1 - (div / NumatRisk), so in the first event shown it is 1-(1/19) = 0.9473684 Note also the following: While NumatRisk continues to fall, with an event or as censored data; when there is an event (divorce), then censor is NA; when there is censored data, survival probability stays the same.

```
plot(DivorceNoCounsel$Years,
      DivorceNoCounsel$SurvivalProbability,
      type = "s", col = "black", xlab = "Years",
      ylab = "Probability of Staying Married")
```

```
lines(DivorceCounsel$Years,
    DivorceCounsel$SurvivalProbability,
    type = "s", col = "black", lwd = 4)
legend("bottomleft",
    c("No counselling", "Counselling"),
    lwd=c(1,4))
```

First, we read the datasets into the R environment using the read.csv() function and assign the dataset names (DivorceCounsel for the datasets containing subjects who received marriage counselling and DivorceNoCounsel for the datasets containing subjects who did not receive marriage counselling). Then, we use the *plot*() and *lines*() functions to plot the marriage-survival probabilities of the subjects in these two datasets, similar to the chart for the life table method. For the entries of the plot() and lines() functions, the x- and y-variables are the Years and SurvivalProbability variables within each dataset. The plot() function produces the survival graph for couples who did not receive marriage counselling while the lines() function plots the graph for those who did. Both graphs are plotted as step plots (type="s") with black color (col = "black"). Line width for those who received counselling is set at 4 (lwd = 4), as against the default width of 1. As previously, the legend is created to match.

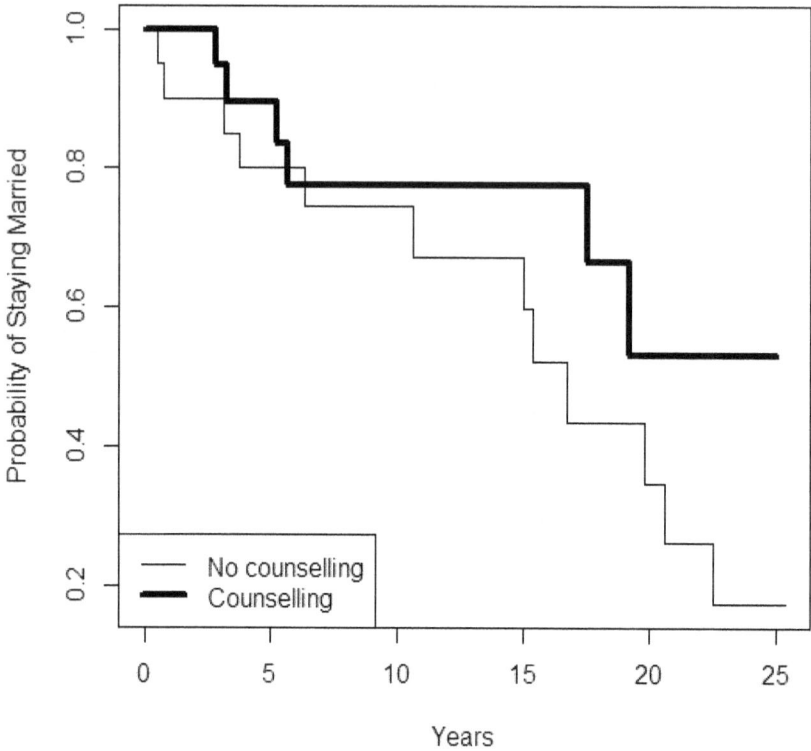

The Kaplan-Meier probability plot can be interpreted in the same manner as the Life Table probability plot. If we want to know the median survival time of the couples' marriages, it's approximately 22 and 15 years for couples who received or did not receive marriage counselling respectively. If we want to look at the proportion of the couples whose marriages lasted longer than 20 years, it's approximately 0.54 and 0.35 for years for couples who received or did not receive marriage counselling respectively.

Confidence limits for the Kaplan-Meier curve

At this point, most of you have probably noticed that the survival probabilities calculated by the Kaplan-Meier method are survival probabilities at a specific time point, which serves as the best guess of the actual survival probability values. In order for us to calculate a range of survival values so that we can be confident that the actual survival values lie within that range, we need to calculate the standard errors (SEs) and 95% confidence intervals of these survival probability values. The calculation of the SEs can be created with **Greenwood's formula** (Collett, 2003; Greenwood, 1926) which is used to calculate the upper and lower limits of the confidence intervals of the survival probabilities. The idea of using confidence intervals is that if we were to resample the same dataset 100 times, there is a 95% chance that the value of the actual survival probability at each time point is within the calculated interval.

To demonstrate the calculation of standard errors and confidence interval estimates of survival probabilities in an example, we can use the dataset containing subjects who did not receive marriage counselling, DivorceNoCounsel. We start by reading the Divorce_NoCounsel.csv file into the R environment using the read.csv() function and assign it the name DivorceNoCounsel.

```
DivorceNoCounsel <-
      read.csv("Divorce_NoCounsel.csv")
```

However, the dataset only has six columns (Years, Counsel, NumAtRisk, div, censor, and SurvivalProbability). We need

to create a few columns for the purposes of running calculations, and then assign the result to a 'Greenwood' column,
which will be *DivorceNoCounsel$Greenwood*.

```
DivorceNoCounsel$Greenwood <-
    DivorceNoCounsel$div /
    (DivorceNoCounsel$NumAtRisk**2 -
    DivorceNoCounsel$NumAtRisk*
    DivorceNoCounsel$div)
DivorceNoCounsel$Greenwood[1] <- 0
```

So we build a new column, DivorceNoCounsel$Greenwood.
The exponent (**) and other operators need to be copied
precisely. (The assignment of zero to the first value of
DivorceNoCounsel$Greenwood is because the first value
has to be $t = 0$, because there were neither 'deaths' – or in
this case, divorces – nor censored data.)

```
GreenwoodSum <- function(var){
   GW <- list()
   for (t in 1:length(var)){
    if (t<2){
      GW <- append(GW, var[t])
    }
    else{
      GW <- append(GW, sum(var[1:t]))
    }
   }
   return(GW)
}
```

We have created a function, *GreenwoodSum()*, which we run by passing the name of a column through its parentheses(). For those unfamiliar with programming, a function performs a set of calculations upon data within one or more arguments: function(data) or function(data, data), etc.

```
DivorceNoCounsel$GreenwoodSum <- unlist(
    GreenwoodSum(DivorceNoCounsel$Greenwood) )
```

In our example, the column DivorceNoCounsel$Greenwood is passed through the GreenwoodSum() function. Then that expression is passed through yet another function, *unlist*(), which converts a list data structure into a vector, a simple structure which contains a sequence of similar elements.

The last expression having created the Greenwood values, we can then create three more columns of values. The last two columns contain the upper and lower confidence interval values:

```
DivorceNoCounsel$SE_SurvivalProbability <-
    DivorceNoCounsel$SurvivalProbability*
    sqrt(DivorceNoCounsel$GreenwoodSum)

DivorceNoCounsel$UL_CI <-
    DivorceNoCounsel$SurvivalProbability +
    1.96*DivorceNoCounsel$SE_SurvivalProbability

DivorceNoCounsel$LL_CI <-
    DivorceNoCounsel$SurvivalProbability -
    1.96*DivorceNoCounsel$SE_SurvivalProbability
```

Now that we have calculated the upper and lower 95% confidence intervals of the survival probabilities, we can plot them to see what they look like graphically. First, we plot the marriage survival probability of the couples who did not receive marriage counselling. We can take the plot() function, enter the Years column of DivorceNoCounsel for the y-axis and SurvivalProbability as the x-axis. We set this plot as type = "s", col = "black", and lwd = 4 so the survival probability plot will be bolder than the confidence limit plots. Then we use the lines() function for the upper and lower 95% confidence limit plots. We assign the UL_CI and LL_CI columns as the y-axes for each lines() function for the upper and lower confidence limit plots respectively.

```
plot(DivorceNoCounsel$Years,
   DivorceNoCounsel$SurvivalProbability,
   type = "s", col = "black", xlab = "Years",
   lwd=4, ylab = "Probability of Couples
   Who Did Not Receive Marriage Counselling
   Staying Married")
lines(DivorceNoCounsel$Years,
   DivorceNoCounsel$UL_CI,
    type = "s", col = "black")
lines(DivorceNoCounsel$Years,
   DivorceNoCounsel$LL_CI,
    type = "s", col = "black")
```

If we run the lines() functions one line at a time, we can see that the plot for upper limit appears above the survival probability plot and the lower limit appears below it. What this plot means is that if we collected the marriage-survival probability data from 100 samples similar to the dataset of DivorceNoCounsel in the population, 95% of the survival probability from the 100 samples would fall between the upper and lower limits of the graph (or in other words, their survival probability plots will be within the area between the two upper and lower limit plots). If you like to practice calculating your own 95% confidence interval of the survival probabilities, you can try to repeat what was done in this example using the DivorceCounsel dataset for the couples who did receive counselling.

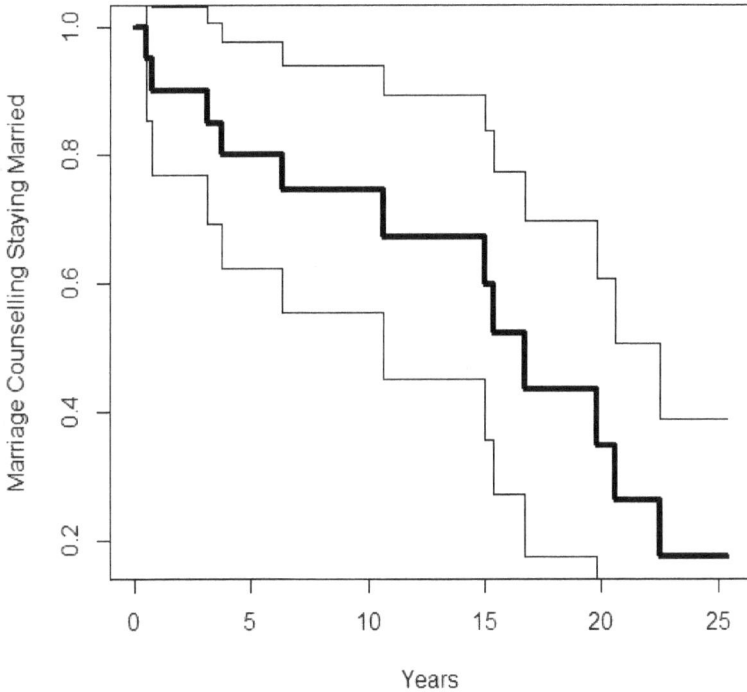

Comparing Survival Curves

Although we can see that the couples who received marriage counselling had an overall higher survival probability than the couples who did not receive marriage counselling, there is a risk of inaccuracy in this observation because it is performed by the naked eye. There is a way to check the statistical significance of the difference between the marriage survival probabilities of these two groups of couples, using the **log-rank test** (Bland & Altman, 2004). The log-rank test is used to compare the degree of overlap between survival plots of two different groups of people in a survival sample, thereby determining whether the marriage survival probability of the couples are significantly different. In this example, the null hypothesis (H_0) is "There is no difference in marriage survival probability between the couples who received or did not receive counselling at any point." The alternative hypothesis (H_A) is "There is a difference in marriage survival probability between the couples who received or did not receive counselling in at least one time point."

The log-rank statistic's distribution is similar to a chi-square statistic distribution, therefore we use the chi-square symbol (χ^2) to represent it and use the chi-square table to determine the significance of the calculated value. The R package *survival* provides a calculation of the log-rank statistic, which will use just a few lines of code!

Before we conduct the log-rank test, we must change our survival dataset (in our case, two separate datasets) so it is in the proper format for the log-rank test. For a dataset to be suitable for the log-rank test from the survival package, the dataset needs to have three columns: 1) a column containing the survival time for each individual (in our case, the marriage survival time column Years in each of DivorceCounsel and DivorceNoCounsel), 2) a status column that represents whether or not there is an event (in this case, a divorce) for the individual at a timepoint, and 3) a group column (in our case, the Counsel column) that represents which group the individual in the survival dataset belongs to.

```
DivorceCounsel <-
    read.csv("Divorce_Counsel.csv")
DivorceNoCounsel <-
    read.csv("Divorce_NoCounsel.csv")

DivorceCounselOrNoCounsel <-
    rbind(DivorceCounsel[2:21,],
    DivorceNoCounsel[2:21,])
```

Since the datasets DivorceCounsel and DivorceNoCounsel both have each of these three columns, with identical column names and column order, we can bind the two datasets together by the rows. To do this, we can use the *rbind()* function and place DivorceCounsel[2:21,] and DivorceNoCounsel[2:21,] inside of it, separated by a comma. We only bind the rows 2 to 21 of each dataset because for both datasets, the first rows had value 0 for the marriage survival time, which is not needed for a log-rank test. We assign the resulting dataset the name *DivorceCounselOrNoCounsel.*

Then we need to install the R package *survival* and use the library() function to open the functions in the package. To perform the log-rank test, we need the Surv() and survdiff() functions in the survival package.

```
install.packages("survival")
library(survival)

div_LogRank<-
      survdiff(Surv(Years, div) ~ Counsel,
      data = DivorceCounselOrNoCounsel)
```

The *survdiff()* function tests for the difference between two more survival curves. Inside this we see the *Surv()* function, which creates a 'survival object' from the time variable (here, 'Years' as the marriage survival time, and the status ('div' as divorce or no divorce), interacting with whether or not the couples received counselling (~ Counsel). To the right is the name of the dataset being used. Everything is assigned to div_LogRank. To look at the result of the log-rank test, we can simply run the name div_LogRank.

```
div_LogRank
```

```
Call:
survdiff(formula = Surv(Years, div) ~ Counsel, data = DivorceCounselOrNoCounsel)

            N Observed Expected (O-E)^2/E (O-E)^2/V
Counsel=0  20       12     8.72      1.23      2.41
Counsel=1  20        6     9.28      1.16      2.41

 Chisq= 2.4  on 1 degrees of freedom, p= 0.1
```

From the output, we can see that the p value is 0.1. Since 0.1 is greater than 0.05, we can conclude that there is not a significant difference in the Kaplan-Meier curves for the marriage survival times between the couples who received or did not receive counselling at the 95% confidence level.

A semi-parametric survival method: Cox Model

Now that we have taken a look at the log-rank test and how it can be used to look at how one factor (receiving or not receiving marriage counselling) can influence the survival time and status, you might ask "April, what if I want to look at multiple factors and how they can influence the survival time and status of something similar to marriage survival?" This question leads to one of the most popular methods in survival analysis, called the **Cox proportional hazards model**. This is often called a **Cox model** or **Cox regression** for simplicity (Cox and Oakes, 1984). When we fit a Cox proportional hazards model to a set of data, we need to make sure that four assumptions are met:

1) **Proportional hazards** – This is the most important assumption for the Cox-proportional hazards model, so important as to feature in the name! What this means is that the ratio of the hazards between the individuals in the dataset must be a constant over time. The way we can tell that the hazards are not proportional is by looking at the Kaplan-Meier Curves. If the curves cross each other or if one curve levels off while the other drops to zero, then hazards are not proportional. Returning to the marriage survival example,

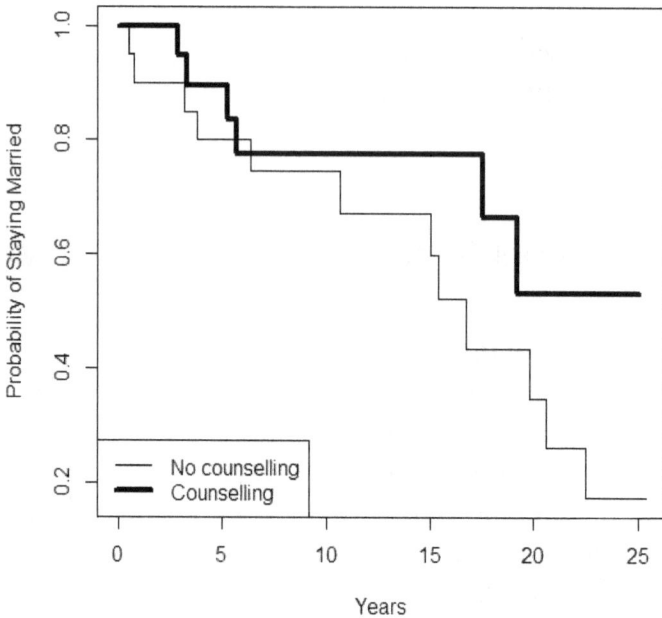

we can see that this is an example of non-proportional hazards because the curves cross each other and one curve levels off while the other drops to zero. The proportional hazards assumption can be tested using **Schoenfeld residuals**.

2) **Linearity** – It is assumed that covariates that are continuous have a linear relationship with the outcome of the Cox model. We can use **Martingale residuals** to check for nonlinearity of continuous covariates. Nonlinearity is not an issue for categorical covariates.

3) **Outliers and other influential values** – Like regression models, we need to check for outliers and other influential values that could build a model that does not

reflect the overall pattern of the data. To check for them, we can use **deviance residuals** or **dfbeta values**.

4) **Independence** – Also as in regression, the data collected are assumed to be independent. This assumption is something that should be controlled for at the data collection level and can make the difference between a valid and an invalid model.

Unlike a regular linear regression, where we need to check the assumptions before building the model, for Cox models we build the theorized model first and then test the assumptions in order to determine the validity of the model. In this example, we will use the ECG dataset, comprising patients who suffered from heart attacks in the past, their survival time, age when the heart attack happened, and other information related to heart health.

```
ECG <- read.csv("ECG.csv")
install.packages(c("survival", "survminer"))
library(survival)
library(survminer)
```

To start, we read the dataset into the R environment using read.csv() and name the dataframe ECG. To perform a Cox regression, we need two R packages, *survival* and *survminer*, here installed at the same time by using the c() function inside the install.packages() function.

```
head(ECG)
```

```
   survival alive age pericardialeffusion contractility heartsize sex
1        11     0  71                    0         9.000     4.600   1
2        19     0  72                    0         6.000     4.100   1
3        16     0  55                    0         4.000     3.420   1
4        57     0  60                    0        12.062     4.603   1
5        19     1  57                    0        22.000     5.750   1
6        26     0  68                    0         5.000     4.310   1
```

```
# Building the Cox Proportional Hazard Model
ecg.cox <-
    coxph(Surv(survival, alive) ~ age
    + heartsize, data = ECG)
```

To build the Cox model, we need the *coxph*() function from the 'survival' package. Inside the coxph() function, we place the specifications for the model and the dataset used to build it. 'Surv' is a function in the survival package; inside it, we place the names of the variables in the ECG dataset representing the survival time and survival status of each individual. On the right of the Surv() function we place two further variables, the age of the individuals at the time of heart attack and their current heart sizes. We then specify ECG as the dataset. We name the model 'ecg.cox'.

Before we go into the model, it is important to first understand that the Cox-proportional hazard model contains a **baseline hazard** function; put crudely, this means that all covariates are equal to zero. This is then compared with a proportional hazards condition, where covariates have a multiplicative relationship with the hazard. As will become apparent when we look at the output, the results may be treated in a similar fashion to those of a multiple regression.

The method for building the Cox-proportional hazard model, selecting the appropriate independent variables for the model, is the same as when you build a logistic or linear regression. You can use forward addition and backward elimination methods for independent variable selection.

The next expression generates output showing the characteristics of the model:

```
summary(ecg.cox)

call:
coxph(formula = Surv(survival, alive) ~ age + heartsize, data = ECG)

  n= 89, number of events= 31

               coef exp(coef) se(coef)     z Pr(>|z|)
age         0.05721   1.05888  0.02088 2.740  0.00614 **
heartsize   0.61636   1.85217  0.22340 2.759  0.00580 **
---
Signif. codes:  0 '***' 0.001 '**' 0.01 '*' 0.05 '.' 0.1 ' ' 1

            exp(coef) exp(-coef) lower .95 upper .95
age             1.059     0.9444     1.016     1.103
heartsize       1.852     0.5399     1.195     2.870

Concordance= 0.696  (se = 0.047 )
Likelihood ratio test= 15.51  on 2 df,   p=4e-04
wald test             = 14.61  on 2 df,   p=7e-04
Score (logrank) test = 15.1  on 2 df,    p=5e-04
```

From the output, first we can look at the likelihood ratio, Wald, and score (log-rank) tests. They are the tests for the significance of the model. For all three tests, the p values are below 0.05, which means that the model is significant. The first table contains three columns that are important for interpreting the model: 1) the coefficients of the independent variables age and heartsize, 2) the exponential value of the negative versions of the coefficient values of the independent variables age and heartsize, 3) the p value of the independent variables. From the p values, which are both less than 0.05, we can see that age and heartsize both significantly influence the hazard. The coefficients of age and heartsize are 0.05721 and 0.61636 respectively, and by taking the exponent of these two values, the effects of these two variables are 1.05888 and 1.85217. The exponential values 1.05888 and 1.85217 are called **hazard ratios**, which are interpreted as follows:

if HR = 1: the independent variable has no effect on the outcome hazard

if HR < 1: the effect of the independent variable leads to a reduction in the outcome hazard

if HR > 1: the effect of the independent variable leads to an increase in the outcome hazard

We can next calculate the increase (or reduction) of the outcome hazard for variables, by subtracting 1 from the hazard ratios and multiplying by 100. So for age, we have (1.059 - 1) = 0.059, and then 0.059 x 100 = 5.9; for the variable heartsize, we have (put more formally) (1.852 -1) x 100 = 85.2. So the age and heart size of the individuals, as they increase by 1, both lead to an increase in the outcome hazard (or death) for the individuals in the dataset by 5.89% and 85.2% respectively. The table below the model information table contains the 95% confidence intervals of the coefficients of the model. For age, the 95% CI is (1.016, 1.103) while for heartsize, the 95% CI is (1.195, 2.870).

```
# Test for Proportional Hazards
test_PH <- cox.zph(ecg.cox)
test_PH
```

```
              chisq df    p
age         0.5586  1 0.45
heartsize  0.0365  1 0.85
GLOBAL     0.5988  2 0.74
```

Now that we have our Cox model built, we can test for the assumptions. We will start by testing the proportional hazards assumption. To do this, we can use the *cox.zph*() function from the survival package on the model we built as ecg.cox. We call the result of this test "test_PH" and look at the output by running test_PH by itself. From the output, we can see that the independent variables age and heartsize as well as the global test all have p values higher than 0.05 (which means they are insignificant). Therefore, we can conclude that there is no significant evidence that the proportionality assumption is violated.

```
ggcoxzph(test_PH)
```

Also, as previously mentioned, we can check for violation of the proportional hazard assumption by looking at the plots of the Schoenfeld residuals against a transformed time variable. For this, we can use the *ggcoxzph*() function from the survminer package. This allows us to diagnose a violation of this assumption if there is a relationship between the Schoenfeld residuals (the round dots in the graphs) and time. We can see that there is no relationship (no increase or decrease) of the values of the Schoenfeld residuals as time progresses. Therefore, we can assume that the proportional hazard assumption is not violated for this model.

Global Schoenfeld Test p: 0.7413

Schoenfeld Individual Test p: 0.4548

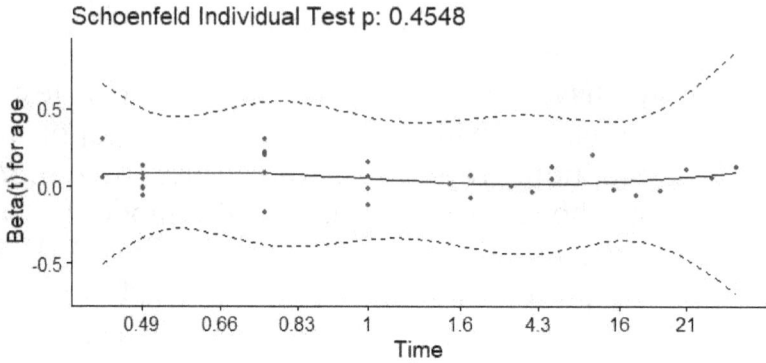

Schoenfeld Individual Test p: 0.8485

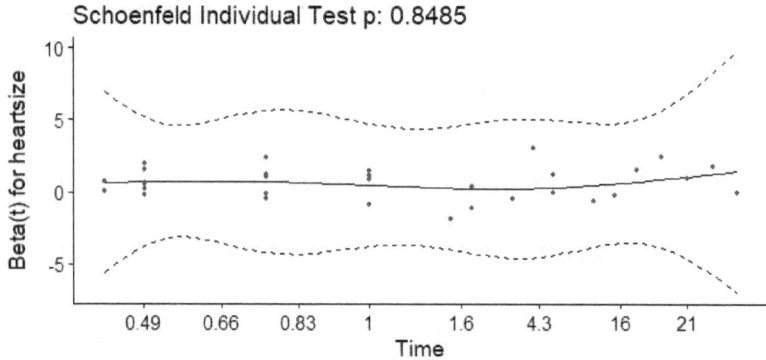

Although in this example the proportional hazard assumption is not violated, in real life, we are rarely this lucky. With a dataset where the proportional hazard assumption is violated, there are two frequently used methods that allow us to still analyze the data using Cox-Proportional hazard model. These two methods involve 1) adding an interaction variable of the variable that violates the proportional hazard assumption (with p value > 0.05) and the time variable (called the time-dependent variable), and 2) stratification.

Time dependent variables and stratification are beyond the scope of this book. External resources you can consult to understand these topics include Fox and Weisberg (2019).

Now that we have tested the proportional hazard assumption, we can now test for the outlier (influential observation) assumption. To test this assumption, we need the *ggcoxdiagnostics*() function from the survminer package. We can check for outliers by looking at both the *dfbeta* as well as the deviance residuals plotted against each observation. To do this, we enter *ecg.cox* (the Cox model we built) as the first entry in the ggcoxdiagnostics() function, specify the type of residual as either 'dfbeta' or 'deviance', set linear.predictions = FALSE, and set ggtheme = theme_bw().

```
# Test influential observations
ggcoxdiagnostics(ecg.cox, type = "deviance",
    linear.predictions = FALSE,
    ggtheme = theme_bw())
```

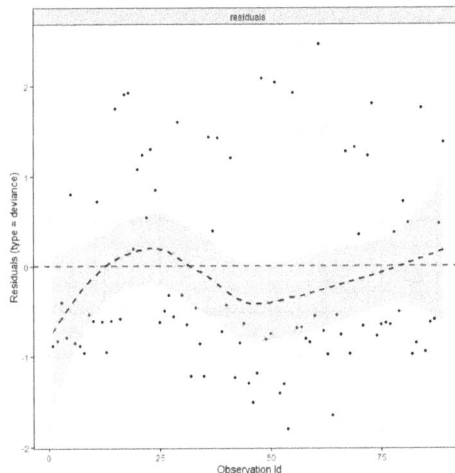

Above is the **influential observation test** using *deviance residuals*. We can see that the pattern of the dots (deviance residual values for each observation) above and below the zero line have relatively the same pattern. There appear to be no influential observations in this dataset.

```
ggcoxdiagnostics(ecg.cox, type = "dfbeta",
    linear.predictions = FALSE,
    ggtheme = theme_bw())
```

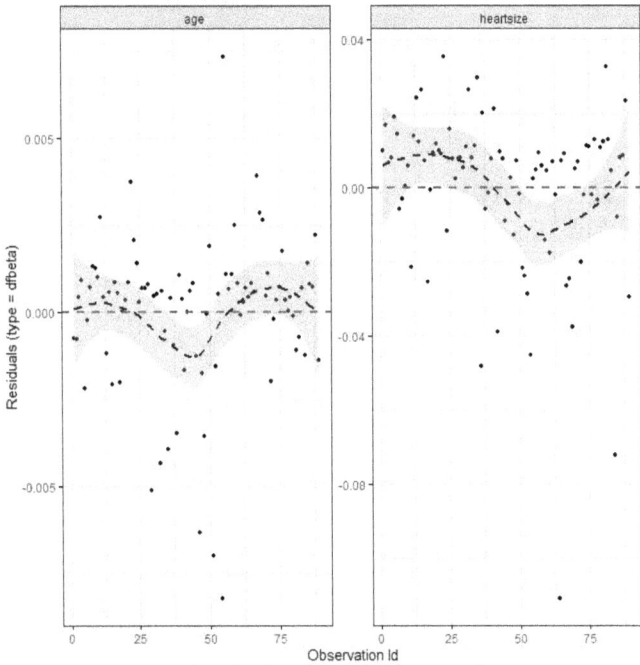

Above is the influential observation test using *dfbeta* residuals. These residuals are essentially changes in the coefficients of the variable if each observation were taken out of the dataset. The graphs show that for both age and heartsize variables, although there are some really large and small values, most of the dfbeta values are still concentrated in the zero line area, so the extreme values did not have a significant influence (or change) on the coefficients of age and heart in the model. The results from the dfbeta residuals and deviance residuals show that there are no outliers in the dataset that need removal or investigation.

Lastly, we need to check the assumption of linearity, by checking for non-linearity. To do this, we use the Martingale residuals of a null Cox model against the values of our model's continuous variables and a couple of their transformation forms to see if the linearity assumption holds. The linearity assumption only applies to continuous (numerical) variables, so even if you have a situation where categorical variables are used for your Cox-proportional hazard model, only the continuous variables need to be tested for this assumption. The ggcoxfunctional() function from the survminer package is used. The variables age and heartsize are included with their log and squareroot transformations:

```
ggcoxfunctional(Surv(survival, alive) ~ age
    + log(age) + sqrt(age) + heartsize
    + log(heartsize) + sqrt(heartsize),
    data = ECG, point.col = "black")
```

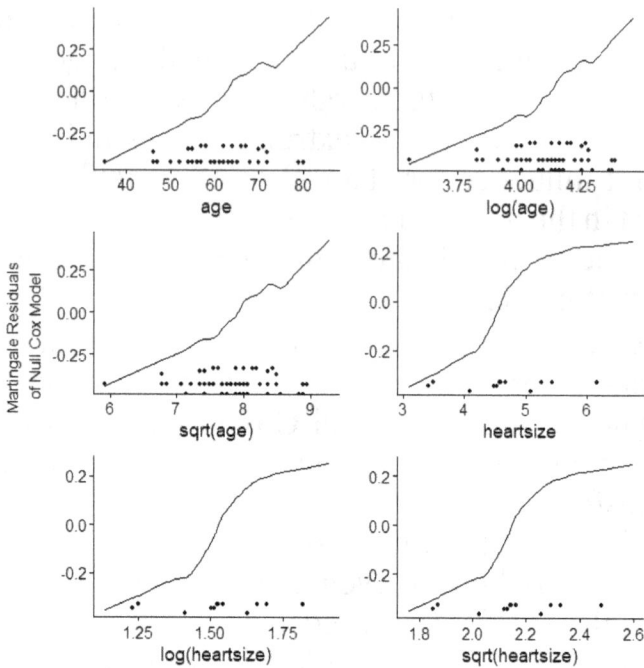

Above is the linearity assumption test for age and heartsize. Each of the six graphs show Martingale residuals plotted on the age, heartsize and their respective log and square root transformations. The black line is called the LOESS function and it is expected to be both straight and horizontal in order for the linearity assumption test to be passed. As we can see from all **6** plots, none of the plots show a straight, horizontal line, so we cannot assume linearity for age and heartsize. These variables will need to be transformed and the Cox-proportional hazard model rebuilt using the transformed versions of these variables that can pass the linearity assumption test. It should be noted that the Cox-proportional hazard model is an iterative process.

Parametric survival methods: Weibull and Exponential Distributions

The Cox proportional hazard model is a semi-parametric model because there are no assumptions about the distribution (shape) of the baseline hazard function, although we do have assumptions for the shape of the independent variables (Collett, 2003). Let's say that we know the shape of the baseline hazard function beforehand, and it fits a shape which we use frequently, then we have a parametric model on our hands (Richards, 2012)! A parametric survival model (also known as parametric proportional hazard model) may not be used very frequently in real life analyses but knowing them can be a great tool if you are in a situation where you are certain that the baseline hazard function takes on a specific distribution and wish to predict the probability of survival to a high degree of accuracy.

At the beginning of this chapter, we cited two frequently used parametric methods, Weibull and exponential. These refer to the most frequently used distributions that the baseline hazard may be assumed to have in a parametric survival analysis. Other distributions do exist, but these two are used most frequently. Before we get into how these methods can be used in parametric survival analysis, let's have a look at the two different distributions.

The **Weibull distribution** is able to handle skew, expecting an increase or decrease over time, which makes it valuable in reliability as well as survival studies. Two parameters determine its appearance: the scale, λ (called lambda), and the shape, k, both of which can have values from zero upwards. The scale determines how flat or heightened the distribution curve may be. The shape determines an increase or decrease of hazard rate:

If k > 1, it means that the hazard rate is increasing over time. In a graph, it means the values will increase, then decrease to create a bump (when the k value is bigger, it means the bump in the graph will also be sharper than a graph with a lower k that is above 1).

If k < 1, it means that the hazard rate is decreasing over time. In a graph, the values decrease quickly over time.

If k = 1, this means that the hazard rate is constant with time. In a graph, it means that the values in the graph are decreasing over time, but the decreasing is slow and constant.

Also when k = 1, this special condition of a Weibull distribution is the same as an **exponential distribution**! In the exponential distribution, the hazard function is constant and not dependent on time. Using the exponential distribution, the lambda is called 'rate' rather than scale.

In cases when a variable takes on a normal distribution when we take the natural log of every single value in the variable, then the **log-normal** distribution (also known as the **Galton distribution**) may be adopted. This is not covered in this book; it is created in a similar way to Weibull and exponential, but is interpreted somewhat differently.

Now that we have a simple understanding of these distributions, we can start learning how to use them by applying them in an example. The dataset we are going to use for this example is a dataset that contains the survival data of patients with pancreatic cancer who received surgery, and the proportion of the highest dose for different drugs.

```
PancreaticCancer <-
    read.csv("PancreaticCancer.csv")
head(PancreaticCancer)

  survival alive drug1 drug2 times_followup surgery
1 2.881709     1     3    44              5       1
2 3.983858     1    32    53              6       0
3 2.855783     0    57    99              5       0
4 8.432354     0    53    75              5       0
5 1.754650     0    87    97              6       0
6 4.142536     1    61    87              7       1
```

First, we read the dataset into the R environment using the read.csv() function and name the dataset PancreaticCancer. The survival time is the number of months each patient lived after diagnosis, the drug variables contain the percentage of the highest dose of drugs used to treat pancreatic cancer. The dataset also contains the number of follow-up appointments patients had with physicians after diagnosis as well as whether or not they received surgery for the disease. In the following example, we will only consider the 'drug1' and 'surgery' variables as independent variables for analysis.

```
library(survival)
Weibull <- survreg(Surv(survival, alive)
            ~ drug1 + surgery,
    data = PancreaticCancer, dist="weibull")
summary(Weibull)
```

We can start by building the Weibull model. To do this, we use the *survreg()* function from the survival package. We first enter the model with Surv(survival, alive) as outcome and drug 1 and surgery as independent variables, with PancreaticCancer as data, specifying with 'dist' the Weibull distribution. We name the model Weibull. To get and interpret the output, we apply the summary() function to the object called Weibull.

```
call:
survreg(formula = Surv(survival, alive) ~ drug1 + surgery, data = PancreaticCancer,
    dist = "weibull")
              Value Std. Error     z       p
(Intercept)  2.05199    0.11190 18.34 < 2e-16
drug1        0.00605    0.00180  3.36 0.00079
surgery     -0.24300    0.10474 -2.32 0.02034
Log(scale)  -0.71663    0.08034 -8.92 < 2e-16

Scale= 0.488

Weibull distribution
Loglik(model)= -281.4   Loglik(intercept only)= -289
        Chisq= 15.15 on 2 degrees of freedom, p= 0.00051
Number of Newton-Raphson Iterations: 7
n= 200
```

From the output, we can see that the loglikelihood test has a p value of 0.00051 (below 0.05), which means that the model with the coefficients drug1 and surgery are significantly better than a model with no variables. We can see that both drug1 and surgery had p values < 0.05, which means that both variables are significant and can be kept in the model. If we want to interpret the other outputs of this model, we need to rebuild the model using another function, *flexsurvreg()* from the *flexsurv* package.

```
install.packages("flexsurv")
library(flexsurv)
Weibull2 <- flexsurvreg(Surv(survival, alive)
```

```
                  ~ drug1 + surgery,
     data = PancreaticCancer, dist="weibullPH")
```

Weibull2

We install and open the package as usual, then placing the model and data into the flexsurvreg function, naming the new model Weibull2. We specify the distribution as "weibullPH" (Weibull Proportional Hazards) to get the outputs for a Weibull model.

```
Call:
flexsurvreg(formula = Surv(survival, alive) ~ drug1 + surgery,
    data = PancreaticCancer, dist = "weibullPH")

Estimates:
         data mean   est       L95%      U95%      se        exp(est)  L95%      U95%
shape    NA         2.04751   1.74920   2.39669   0.16450   NA        NA        NA
scale    NA         0.01497   0.00683   0.03281   0.00599   NA        NA        NA
drug1    50.22500  -0.01240  -0.01965  -0.00514   0.00370   0.98768   0.98054   0.99487
surgery   0.54000   0.49755   0.07272   0.92239   0.21676   1.64469   1.07543   2.51529

N = 200,  Events: 89,  Censored: 111
Total time at risk: 1100.922
Log-likelihood = -281.4432, df = 4
AIC = 570.8865
```

From the output, we can see that the coefficients of these two variables are -0.01240 and 0.49755. To make these coefficients relevant to the interpretation, we need to take the exponential of these values and then subtract 1 from the value, then multiplying by 100. For drug1, the variable is the percentage of the maximum dose of a drug that's given to a patient for treatment: the exponential is 0.98768, so 0.98768 -1 = -0.01232; multiply by 100 and this becomes -1.2. This means that if we increase the percentage of the drug by 1%, then the patient will have 1.2% reduction of hazard than if otherwise. For the variable surgery, which represents whether or not a patient received surgery for their cancer, the exponential is 1.64469, so subtracting 1, we have .64469; multiplied by 100, this is 64.5. This means that for a person who received surgery (variable coded as 1 = received surgery, 0 = not received surgery) will have 64.5% increase in hazard than a person who did not receive it. The scale and the shape of the distribution of the model are 0.01497 and 2.04751 respectively. The shape being higher than 1 means there will be a bump in the graph. The scale being 0.01497 is quite low which means the graph will be quite flat.

The exponential model will take on the same idea as the Weibull, except instead of the dist specification being "weibullPH", we put it as "exponential". We first build the model using the survreg() function from survival and name the model Exponential and use the summary() function to obtain the output.

```
Exponential <- survreg(Surv(survival, alive)
              ~ drug1 + surgery,
    data = PancreaticCancer, dist="exponential")
summary(Exponential)
```

```
call:
survreg(formula = Surv(survival, alive) ~ drug1 + surgery, data = PancreaticCancer,
    dist = "exponential")
              Value Std. Error    z      p
(Intercept)  2.24477   0.23284   9.64  <2e-16
drug1        0.00831   0.00362   2.29   0.022
surgery     -0.28620   0.21367  -1.34   0.180

Scale fixed at 1

Exponential distribution
Loglik(model)= -309.4   Loglik(intercept only)= -312.9
        chisq= 6.92 on 2 degrees of freedom, p= 0.032
Number of Newton-Raphson Iterations: 4
n= 200
```

From the output, we can see that the exponential model is different from the Weibull model in the sense that the surgery variable is no longer significant (with p value > 0.05). However, from the loglikelihood test, we can see that the p value is 0.032, which is lower than 0.05. Therefore, the model is still acceptable under the assumption of an exponential distribution for the baseline hazard, but the surgery variable does not have a significant influence on the outcome hazard. In your own analysis, it is ok to remove an insignificant variable from a model like this or change it to see if there is a way to include it in the model, but for this example, we will keep it there.

To interpret the results from the exponential model, we use the flexsurvreg() function from the flexsurv package to rebuild the model and call it Exponential2:

```
Exponential2 <-
       flexsurvreg(Surv(survival, alive)
             ~ drug1 + surgery,
data = PancreaticCancer, dist="exponential")


Exponential2
```

```
Call:
flexsurvreg(formula = Surv(survival, alive) ~ drug1 + surgery,
      data = PancreaticCancer, dist = "exponential")

Estimates:
         data mean   est        L95%       U95%       se         exp(est)   L95%       U95%
rate         NA       0.10595    0.06713    0.16721    0.02467         NA        NA         NA
drug1    50.22500    -0.00831   -0.01540   -0.00121    0.00362    0.99173    0.98472    0.99879
surgery   0.54000     0.28620   -0.13259    0.70498    0.21367    1.33136    0.87583    2.02381

N = 200,  Events: 89,  Censored: 111
Total time at risk: 1100.922
Log-likelihood = -309.4012, df = 3
AIC = 624.8023
```

From here, we can see that the lambda value ('rate') is 0.10595. This means the graph is likely to be very flat. Also, for drug 1 and surgery, the coefficients are -0.00831 and 0.28620 respectively. For drug1, the results show that if we increase the percentage of the drug by 1%, then the patient will have 0.8% increase in hazard (1 - the exponential, meaning 1 - 0.99173 = 0.00827; then multiply by 100) than if otherwise. For surgery, if a person received surgery, the person will have 33% (1.33136-1 = 0.33136, then multiply by 100) decrease in hazard than a person who did not.

Summary

The methods for survival analysis (under the 3 branches of non-parametric, semi-parametric, and parametric) can seem like they are very different from each other, but this is hardly the case. If you find a survival dataset to analyze, as long as it contains the survival time, the survival condition (alive, not alive), and variables that could influence the survival of the subjects, it's likely that all of the methods under the survival analysis umbrella can be applied to the dataset. If you know the distribution of the baseline hazard of the survival data and want to look at the influence of multiple variables on the survival/hazard outcome, consider using parametric methods. If you don't know the distribution of the baseline hazard of the survival data and still want to look at the influence of multiple variables, then consider a semi-parametric method like the Cox-proportional hazard model. If you don't know the distribution of the baseline hazard and only want to consider one variable's influence on the outcome, consider methods under the non-parametric branch. It's time for you to find your own survival dataset and perform analysis of your choices using these methods!

Further reading

If you would like to learn about the survival analysis methods in more detail, *Survival Analysis: A Self-Learning Text* by Kleinbaum and Klein, and *Modelling Survival Data in Medical Research* (2nd ed.) by Collett are great references!

Chapter 4 – Longitudinal Analysis

Sections include

Repeated Measures ANOVA
  one-way
  two-way

Linear Mixed Effects Model

Generalized Estimating Equations (GEE)

Introduction

So far in this book, we have covered analysis methods designed to analyze data with more than one timepoint. Time series analysis methods are used to predict future values in one single variable based on the trends and seasonality of previous values of the same variable. These usually involve only one subject (e.g. weather temperature at one place, price of one stock, etc.) with data collected at many timepoints (up to hundreds, or more) in real life (Shumway and Stoffer, 2017).

The topic of this chapter, longitudinal analysis, is a method that is similar to time series analysis in the sense that it also involves analyzing data collected at multiple timepoints, but the data involves more than one subject and fewer timepoints. **Longitudinal analysis** (aka **panel analysis** or **repeated measures analysis**) is another collection of methods designed to analyze data collected from multiple individuals, but at a few different time points, at least 2, but never as many as time series data (Diggle *et al*, 2002).

Its purpose is very different from time series analysis. The main goal of time series analysis is to predict future values of the variable in question. Longitudinal analysis, however, is designed to examine *differences* between variable values collected at different timepoints between different individuals, and whether or not other variables in the dataset influence values at different time points (Fitzmaurice *et al*, 2004).

There are various methods designed to model longitudinal data. Some you see frequently in studies are **repeated**

measures ANOVA, linear mixed effects, and **generalized estimating equations (GEE)**. Before we go into detail about these three longitudinal data analysis methods, it is best to start by understanding longitudinal data in detail as well as in what circumstances these different methods can be used.

Subjects	Timepoint 1	Timepoint2	Timepoint3
Subject 1	Y(Subject1, Timepoint1)	Y(Subject1, Timepoint2)	Y(Subject1, Timepoint3)
Subject 2	Y(Subject2, Timepoint1)	Y(Subject2, Timepoint2)	Y(Subject2, Timepoint3)
Subject 3	Y(Subject3, Timepoint1)	Y(Subject3, Timepoint2)	Y(Subject3, Timepoint3)
Subject 4	Y(Subject4, Timepoint1)	Y(Subject4, Timepoint2)	Y(Subject4, Timepoint3)

As shown here, longitudinal analysis involves collecting data of the same variable from multiple subjects at multiple timepoints.

In studies, we often want to examine whether the mean values of the variables at each timepoint are different from each other; repeated measures ANOVA is a great method for this purpose (Girden, 1992).

If we want to see the effect of variables in the dataset on the population average of the longitudinal outcome variable, then generalized estimating equations (GEE) should be used (Liang and Zeger, 1986).

To see the effect of the variables in the dataset as well as the random variables (such as differences between each different subject) on the population average of the longitudinal outcome variable, the linear mixed model is the method to use (Laird and Ware, 1982).

Let's go through each method to understand each of them in more detail and how to use them to analyze longitudinal data!

Repeated Measures ANOVA

Repeated measures ANOVA is also called **within-subjects ANOVA**. The term *within-subjects* gives a clear definition of the nature of the method, because it is a method that requires us to perform ANOVA on the means of the outcome values of the *subjects* in the datasets for each timepoint. In order for a repeated measures ANOVA to be correctly performed, the data must meet three assumptions: 1) there are no significant outliers, 2) there is a normal distribution, and 3) it passes the sphericity test, whether or not the differences of the data values between different timepoints are the same (Girden, 1992). Potential outliers may be found with a box plot. The normality assumption can be tested using the Shapiro-Wilk test. Mauchly's test of sphericity accounts for the third assumption. Let's go through an example in R of how to check the assumptions and perform the repeated measures ANOVA.

We will use the file stimulus.csv (data adapted from Davis, 2019). 20 adults with autism, 10 female and 10 male, are introduced to an experimental visual stimulus to see if it has any impact on stereotypical behaviors. Recorded stereotypical behaviors per person are rated with a scale of

1 as the lowest level and 10 as the highest. There are three phases: timepoint 1 (t1) precedes the stimulus; t2 represents the stimulus period; t3 is post-stimulus.

```
file <- read.csv("Stimulus.csv")
file
```

	case	score	timepoint	factor
1	1	6	t1	Female
2	2	4	t1	Female
3	3	9	t1	Female
4	4	7	t1	Female
5	5	6	t1	Female
6	6	7	t1	Female
7	7	5	t1	Female
8	8	6	t1	Female
9	9	4	t1	Female
10	10	6	t1	Female
11	11	8	t1	Male
12	12	5	t1	Male
13	13	8	t1	Male
14	14	4	t1	Male
15	15	7	t1	Male
16	16	8	t1	Male
17	17	5	t1	Male
18	18	8	t1	Male
19	19	3	t1	Male
20	20	9	t1	Male
21	1	4	t2	Female
22	2	3	t2	Female
23	3	6	t2	Female
24	4	7	t2	Female
25	5	6	t2	Female
26	6	5	t2	Female
27	7	4	t2	Female
28	8	4	t2	Female
29	9	3	t2	Female
30	10	4	t2	Female
31	11	5	t2	Male
32	12	4	t2	Male
33	13	5	t2	Male
34	14	6	t2	Male
35	15	6	t2	Male
36	16	7	t2	Male
37	17	3	t2	Male
38	18	5	t2	Male
39	19	4	t2	Male
40	20	6	t2	Male
41	1	6	t3	Female
42	2	4	t3	Female
43	3	8	t3	Female
44	4	7	t3	Female
45	5	7	t3	Female
46	6	6	t3	Female
47	7	4	t3	Female
48	8	5	t3	Female
49	9	5	t3	Female
50	10	5	t3	Female
51	11	8	t3	Male
52	12	4	t3	Male
53	13	6	t3	Male
54	14	8	t3	Male
55	15	6	t3	Male
56	16	7	t3	Male
57	17	6	t3	Male
58	18	7	t3	Male
59	19	6	t3	Male
60	20	8	t3	Male

As you can see, this dataset is in the form where each row is one subject at one timepoint. We have separate columns for the timepoints and for the sex. The same 20 individuals are seen in different rows (at different time points) but their sexes between the time points stay the same.

It is important to note that this data structure is required because the R code used in this chapter to analyze longitudinal data can only be used for data in this format. Sometimes we study datasets where we have different columns for outcome scores at each timepoint; we would need to convert the data into the format shown here.

One Way Repeated Measures ANOVA

```
boxplot(score ~ timepoint, data=file,
  main="Testing outlier assumption - 1 way",
  xlab="time points",ylab="behavioral scores")
```

For the moment, we are only interested in the main effect, that of the stimulus affecting stereotypical behavior. The boxplot is to show the stereotypical scores at different time points, testing for outliers. The first argument of this function is score~timepoint, with the 'score' variable on the y-axis and 'timepoint' is on the x-axis. The data=, main=, xlab=, and ylab= arguments specify the dataset name, and labels for the main plot, x-axis, and y-axis respectively (the labels are entered as strings).

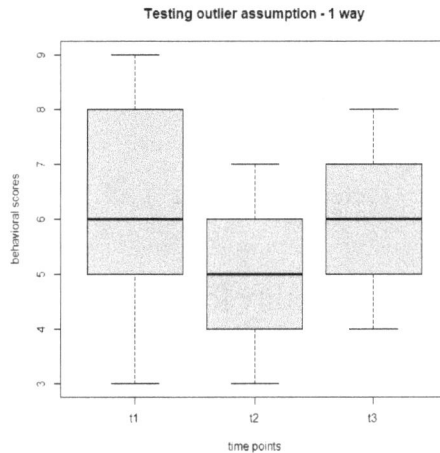

Testing outlier assumption - 1 way

From the boxplot, we can see from the timepoints that there are no data points that are located outside the top and bottom horizontal bars for each of the 3 plots, therefore there are no significant outliers.

```
shapiro.test(file$score [file$timepoint=="t1"])
```

Testing for normality, the *shapiro.test*() function is applied to the stereotypical behavioral score values at each time-point. In the first case, the score values are taken from the dataset where the timepoint variable value is equal to the string "t1".

```
        Shapiro-wilk normality test
data:  file$score[file$timepoint == "t1"]
W = 0.94847, p-value = 0.3444
```

```
shapiro.test(file$score [file$timepoint=="t2"])
```

```
        Shapiro-wilk normality test
data:  file$score[file$timepoint == "t2"]
W = 0.91429, p-value = 0.07698
```

```
shapiro.test(file$score [file$timepoint=="t3"])
```

```
        Shapiro-wilk normality test
data:  file$score[file$timepoint == "t3"]
W = 0.90936, p-value = 0.06196
```

We can test for the normality for timepoint 2 and timepoint 3 by simply replacing the string with "t2" and "t3" for time-points 2 and 3 respectively. From the p values of the outputs of the normality test for the three timepoints, we can see that all of the p values are above 0.05. This means we can reject the null hypothesis that the stereotypical behavior scores at those three timepoints are not normal, which means that the normality assumption is met.

```
install.packages("rstatix") # installs package
library(rstatix) # opens package
RM_ANOVA <- anova_test(data = file,
  dv = score, wid = case, within = timepoint)
```

The *anova_test*() function from the *rstatix* package tests for sphericity and performs the repeated measures ANOVA.

```
options(scipen = 999)
# removes scientific notation
RM_ANOVA
```

```
ANOVA Table (type III tests)

$ANOVA
      Effect DFn DFd     F        p p<.05   ges
1 timepoint   2  38 14.368 0.0000225    * 0.163

$`Mauchly's Test for Sphericity`
      Effect     W     p p<.05
1 timepoint 0.802 0.137

$`Sphericity Corrections`
      Effect   GGe     DF[GG]    p[GG] p[GG]<.05   HFe     DF[HF]    p[HF] p[HF]<.05
1 timepoint 0.835 1.67, 31.72 0.000084         * 0.906 1.81, 34.41 0.0000478         *
```

Let us first look at the results of Mauchly's test for sphericity. The p value is 0.13. As this is higher than 0.05, we can reject the null hypothesis that the differences between the behavioral scores between timepoints are not equal, which means the sphericity assumption test is passed. This means that we can accept the results of the one way repeated measures ANOVA test.

The p value of the repeated measures ANOVA is well below 0.05, so we can reject the null hypothesis, that the means of the stereotypical behavioral scores between the timepoints are not significantly different from each other. From this result, we can deduce that there was a change in stereotypical behavior during the three timepoints; the previous chart showed the major change to be during the period in which the stimulus was shown.

Two Way Repeated Measures ANOVA

The purpose of the *one way* repeated measures ANOVA is to find out if the longitudinal outcome scores are different between the timepoints, by looking at whether the means are significantly different between the timepoints. The *two way* repeated measures ANOVA also looks at whether or not a variable that can potentially divide the subjects in the dataset into separate groups (e.g. their sex), has an impact on the differences of the longitudinal outcome scores at different timepoints. The two way repeated measures ANOVA process is very similar to the one way variant in the sense that it must meet the same three assumptions and that the ANOVA is performed in a similar way.

```
boxplot(score~timepoint + factor, data=file,
   main="Testing outlier assumption - 2 way",
   xlab="time points",
   ylab="stereotypical behaviors")
```

Testing for outliers, we use the boxplot() function again. The first entry has the same score~timepoint, but additionally we add the sex variable (in the file, this is 'factor') because we need to look at the score at each timepoint for outliers in males and females. The 'main' entry is a label, this time stipulating '2–way'. From the output boxplot figure, we can see that there are no points that are outside of the top and bottom lines of the behavioral scores at each timepoint for males and females. So the outliers assumption is met.

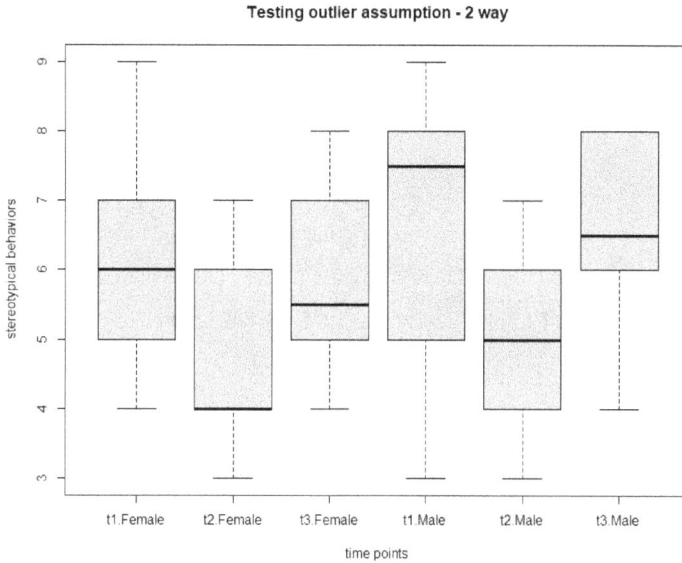

Testing outlier assumption - 2 way

For the normality test, as in the one way ANOVA, we use the Shapiro-Wilk test, running the shapiro.test() function. However, for two way ANOVA, we need rather more code in order to check the normality of stereotypical behavior scores for all 3 timepoints for males and females. Therefore, for each test, the shapiro.test() function requires the score values of the Stimulus data where the timepoint variable is equal to "t1" (or "t2" or "t3") and with "factor", the sex variable, equal to "Female" (or "Male"). As you can see, the p values of all 6 tests are way above 0.05, therefore, the normality assumption is met.

```
shapiro.test(file$score[file$timepoint=="t1"
   & file$factor=="Female"])

shapiro.test(file$score[file$timepoint=="t1"
   & file$factor=="Male"])

shapiro.test(file$score[file$timepoint=="t2"
   & file$factor=="Female"])

shapiro.test(file$score[file$timepoint=="t2"
   & file$factor=="Male"])

shapiro.test(file$score[file$timepoint=="t3"
   & file$factor=="Female"])

shapiro.test(file$score[file$timepoint=="t3"
   & file$factor=="Male"])
```

```
        Shapiro-Wilk normality test
data:  file$score[file$timepoint == "t1" & file$factor == "Female"]
W = 0.91771, p-value = 0.3383

...
data:  file$score[file$timepoint == "t1" & file$factor == "Male"]
W = 0.87668, p-value = 0.1195
data:  file$score[file$timepoint == "t2" & file$factor == "Female"]
W = 0.89597, p-value = 0.1977
data:  file$score[file$timepoint == "t2" & file$factor == "Male"]
W = 0.95194, p-value = 0.6915
data:  file$score[file$timepoint == "t3" & file$factor == "Female"]
W = 0.93185, p-value = 0.4664
data:  file$score[file$timepoint == "t3" & file$factor == "Male"]
W = 0.8724, p-value = 0.1066
```

```
# Sphericity Test
# and repeated measures 2-way anova
library(rstatix)
TW_ANOVA <- anova_test(data = file,
  dv = score, wid = case,
  within = timepoint, between = factor)
```

The sphericity test is performed in a similar way to the one way ANOVA. We take the anova_test() function and place the dataset, dependent variable ('score'), and the unique subject identifier variable (here, 'case'). For the within = and between = arguments, we place the timepoint and factor (sex) variables respectively because timepoint is a within-subject variable and sex is a between-subject variable. We include the between-subject variable to see if it has an interaction effect with the time variable on the differences between the behavioral scores between the different time-points. We call the two way ANOVA result 'TW_ANOVA'.

```
TW_ANOVA
```

```
ANOVA Table (type II tests)

$ANOVA
          Effect DFn DFd      F         p p<.05    ges
1         factor   1  18  1.251 0.2780000        0.048
2      timepoint   2  36 13.840 0.0000348     *  0.171
3 factor:timepoint  2  36  0.303 0.7410000        0.004

$`Mauchly's Test for Sphericity`
          Effect    W     p p<.05
1      timepoint 0.79 0.135
2 factor:timepoint 0.79 0.135

$`Sphericity Corrections`
          Effect   GGe  DF[GG]   p[GG] p[GG]<.05 HFe  DF[HF]   p[HF] p[HF]<.05
1      timepoint 0.827 1.65, 29.76 0.000129      * 0.9 1.8, 32.38 0.0000743     *
2 factor:timepoint 0.827 1.65, 29.76 0.700000        0.9 1.8, 32.38 0.7180000
```

Mauchly's test for sphericity has p values of 0.135. As these are larger than 0.05, the sphericity assumption is satisfied.

If this assumption was not met, the repeated measures ANOVA can be adjusted for lack of sphericity by using the Greenhouse-Geisser correction (above, 'GG') (Greenhouse and Geisser, 1959).

As the results were satisfactory for performing a repeated measures ANOVA on this dataset, let's look at the main ANOVA statistics: we can see that only the timepoint variable is shown to have a p value below 0.05, which means that the mean values of the stereotypical behavior scores are only significantly different between timepoints.   Sex ('factor') and the sex-timepoint interaction are not significant, meaning that the mean behavior scores were not different according to sex and the sex and timepoint interaction.   In summary, if we look at the charts, it seems unlikely that the stimulus had much of a lasting effect, and the patterns appear similar for both sexes.

Linear Mixed Effects Model

Repeated measures ANOVA is a great method to use for longitudinal data if our interest is primarily only the difference between the outcomes at different timepoints. If however we wish to look at how covariates in a longitudinal dataset affect the changes in the outcomes at different timepoints, and also take into consideration that sometimes different individuals and their unique attributes can influence the outcome value at different timepoints, the linear mixed effects model (also known as **mixed model** or **multilevel**

model [*]) is a great modeling method that is designed to allow us to do that.

A linear mixed effects model contains two types of effect variables, fixed and random. **Fixed effects** are the population level effects that are expected to behave in similar ways if the individuals in the population samples changed, while **random effects** are expected to behave differently for different individuals in different population samples. Together, fixed effects and random effects model the average trend of the outcome values as well as how the individuals and different time points may cause the outcome values to deviate from the average trend (Laird and Ware, 1982). The linear mixed effects model also contains fixed intercepts which represent the average intercept, while the random intercepts represent the deviations from this average fixed intercept for different individuals in the dataset.

To explain in more depth how the linear mixed effects model works, let's go through an analysis together! We will use an artificial dataset *Grades* containing longitudinal data: the grades of 24 students who followed a healthier diet or their normal diet with tests at 3 different timepoints.

```
Grades <- read.csv("Grades.csv") # read dataset
install.packages("lme4") # install package lme4
library(lme4) # open package

head(Grades); tail(Grades)
```

---

[*]Those seeking to investigate nested structures via multilevel modeling should consult the section on Hierarchical Modeling in Chapter 6.

```
   id diet grades tests        id diet grades tests
1   1  ctr     83    t1    67 19 Diet     87    t3
2   2  ctr     97    t1    68 20 Diet     88    t3
3   3  ctr     93    t1    69 21 Diet     93    t3
4   4  ctr     92    t1    70 22 Diet     95    t3
5   5  ctr     77    t1    71 23 Diet     91    t3
6   6  ctr     72    t1    72 24 Diet     78    t3
```

When we build a linear mixed effects model, what the model looks like depends upon the analysis of interest. In the Grades dataset, the 'tests' and 'diet' variables are fixed effects because regardless of how much we reselect the sample, the study will always have three tests and control versus diet. The random effect is the 'id' variable because the students will vary and some might score better than others on a test regardless of which test or what their diets are like, which can influence the outcomes at the subject level.

For those who are still a little confused about which variables can be used as fixed or random variables, the random variables are always finite and categorical. Therefore, if you have a continuous variable in the dataset you want to consider for a linear mixed effects model, it will always be a fixed effect; if it is categorical and it is a variable that is used to design the study, then it is also a fixed effect; if it is categorical and it has one category for each individual in the study, then it is a random variable. Therefore, for the Grades dataset, the variables 'tests' and 'diet' can serve as fixed effects.

For random effects, we have to consider both the random intercepts and the random slopes. Random intercepts are used when we suspect and want to model the effect that some individuals are just better at taking tests than others in a dataset and those persons' individual intercepts for test scores are a few points higher than the mean test scores for the population. Random slopes are used when we suspect that a fixed effect affects the individuals differently.

The two models below are two different linear mixed effects models built using the Grades dataset. To build the models, we use the *lmer()* function from the lme4 package we previously installed.

```
mod1 <- lmer(grades ~ tests +
  diet + (1|id), data = Grades)
```

For the first model, mod1, we are building a model of the outcome variable grades with 'tests' and 'diet' as fixed effects. Just by entering them as fixed effects, we are implying that we also want it to generate a fixed intercept for us. Next, we can add the random effects: these are added into the model by placing them in parentheses. If we want to enter a random intercept for a random variable, then we enter it in the form of '(1|random variable)'. The long vertical bar is normally found above the backslash (\ ) on your keyboard.

mod1 only checks to see if there is a significant difference between individuals and how they score on tests, which is why only the random intercept for 'id' was included.

```
mod2 <- lmer(grades ~ tests +
  diet + (1|id) + (0+diet|id), data = Grades)
```

For mod2, in addition to the random intercept for 'id', we

are testing to see if 'diet' (whether it's the healthy or control diet) affects the individuals in the dataset differently. For this, we can add a random slope (0+diet|id). We write the random slope this way because the difference in how the diet affects the subjects at the individual level is actually at the id level. As for the 0+diet, the 0 is for the purpose of telling R that we do not want an extra random intercept for this random slope of how diet affects the subjects at the individual level.

mod1

```
Linear mixed model fit by REML ['lmerMod']
Formula: grades ~ tests + diet + (1 | id)
   Data: Grades
REML criterion at convergence: 430.7528
Random effects:
 Groups   Name        Std.Dev.
 id       (Intercept) 8.387
 Residual             3.185
Number of obs: 72, groups:  id, 24
Fixed Effects:
(Intercept)        testst2        testst3       dietDiet
     85.694         -1.958         -4.625          4.194
```

mod2

```
Linear mixed model fit by REML ['lmerMod']
Formula: grades ~ tests + diet + (1 | id) + (0 + d
   Data: Grades
REML criterion at convergence: 430.2497
Random effects:
 Groups   Name        Std.Dev. Corr
 id       (Intercept) 6.185
 id.1     dietctr     6.904
          dietDiet    4.061    -0.31
 Residual             3.185
Number of obs: 72, groups:  id, 24
Fixed Effects:
(Intercept)        testst2        testst3       dietDiet
     85.694         -1.958         -4.625          4.194
```

From the outputs, you have probably noticed that we do not have p values for the fixed or random effects for the linear mixed models we've just built. So how do we know which variables to include in the model and how do we interpret the effects of the fixed and random variables that are retained in the model? We can first narrow down which model is the better model by using the likelihood-ratio test (Luke, 2017), which can easily be done by applying the *anova*() function to the two models.

```
anova(mod1, mod2)
```

```
Data: Grades
Models:
mod1: grades ~ tests + diet + (1 | id)
mod2: grades ~ tests + diet + (1 | id) + (0 + diet | id)
      npar    AIC    BIC  logLik deviance  Chisq Df Pr(>Chisq)
mod1     6 452.98 466.64 -220.49   440.98
mod2     9 458.43 478.92 -220.22   440.43 0.5488  3      0.908
```

Of the two models, we can see that mod1 has a much lower AIC value than mod2. Also, from the *p* value of 0.908 for the likelihood-ratio test, we can reject the null hypothesis that the two models are significantly different from one another. Therefore, since neither one of these two models are better than the other, we should follow the rule of parsimony for model building and choose mod1 as the better model because it has one less term (the random slope).

For interpreting the results from mod1, the intercept under "Fixed Effects" is 85.694, which is the average intercept (the average grade of the population). Under the intercept for random effect for "id", the standard deviation associated with this intercept is 8.387. This means there is a grade difference of around 8.39 compared to the average grade of 85.694 depending on the individual student. For the "Fixed effects", the results show that the students scored worse on the second and third test than the first by 1.958 points and 4.625 respectively. The students who received the healthier diet also scored 4.194 points better than the students who had a normal diet.

```
summary(mod1)
```

```
Linear mixed model fit by REML ['lmerMod']
Formula: grades ~ tests + diet + (1 | id)
   Data: Grades

REML criterion at convergence: 430.8

Scaled residuals:
     Min       1Q   Median       3Q      Max
-1.94307 -0.58692  0.02337  0.52860  1.94231

Random effects:
 Groups   Name        Variance Std.Dev.
 id       (Intercept) 70.34    8.387
 Residual             10.14    3.185
Number of obs: 72, groups:  id, 24

Fixed effects:
            Estimate Std. Error t value
(Intercept)  85.6944     2.5347  33.808
testst2      -1.9583     0.9194  -2.130
testst3      -4.6250     0.9194  -5.030
dietDiet      4.1944     3.5052   1.197

Correlation of Fixed Effects:
         (Intr) tstst2 tstst3
testst2  -0.181
testst3  -0.181  0.500
dietDiet -0.691  0.000  0.000
```

Generalized Estimating Equations (GEE)

The linear mixed effects model seems very complicated because it allows every single individual in the dataset to have their own unique model. But let's say you only care to model the population average and are not interested in the model down to the subject-specific level. Then generalized estimating equations (GEE) may be a better option (Ballinger, 2004). The *geepack* package in R contains a *geeglm*() function that can help us build our models. There are two requirements before we start using GEE: we need to specify the family (or distribution) of the outcome, and also the correlation structure of the repeated measures (how the repeated measures are correlated).

Now that we have a basic understanding of what GEE is and how to apply it, let's use a longitudinal dataset called *workforce*, adapted from the LaborSupply dataset (Croissant, 2022). It contains longitudinal data for 532 individuals including their daily working hours as well as variables that could potentially influence their working hours (e.g. age, number of children, disability status). First we read this dataset into the R environment using the read.csv() function. Then we install and open two R packages that we will use for building and selecting GEE models. The packages are the aforementioned geepack and also *MESS*.

```
workforce <- read.csv("workforce.csv")
install.packages("geepack") # installs
library(geepack) # opens
install.packages("MESS") # installs
```

```
library(MESS)  # opens
```

```
head(workforce)
```

```
  hrs kids age disab id year
1 7.58    2  27     0  1 1979
2 7.75    2  28     0  1 1980
3 7.65    2  29     0  1 1981
4 7.47    2  30     0  1 1982
5 7.20    4  35     0  2 1979
6 6.95    3  37     1  2 1980
```

Starting on our model building, let's say that we suspect that the number of children a working adult has, their age, and whether or not they are disabled, can influence the number of hours they work daily.

```
gee_mod1 <- geeglm(hrs~kids + age + disab,
    data=workforce, id=id,
    corstr="exchangeable", family = "gaussian")
```

First we place the model like this into the geeglm() function. After that, we specify the dataset and id as the variable that identifies each unique individual in the dataset (the variable which in this dataset is also named "id"). Then we do the most important part of building a GEE model: specifying the correlation structure and family arguments. 'corstra' stands for correlation structure, which explains how the outcome data collected at different time points are correlated. For the geeglm() function, we have 5 options for the correlation structure: independence, exchangeable, ar1, unstructured and userdefined. The 'family' argument is used to specify the distribution of the outcome, and we have binomial, gaussian, gamma, poisson, etc. As the *hrs* variable in the workforce dataset is continuous, we can put it as "Gaussian" for the 'family' argument. If the outcome contains 2 categories, "binomial" is the appropriate option for the family argument; if the outcome is extremely skewed, then "gamma" is preferred.

For our first model, we specify the correlation matrix as "exchangeable". We name this model gee_mod1 and use the summary() function to get information about the model.

```
summary(gee_mod1)

Call:
geeglm(formula = hrs ~ kids + age + disab, family = "gaussian"
    data = workforce, id = id, corstr = "exchangeable")

 Coefficients:
             Estimate    Std.err      Wald Pr(>|w|)
(Intercept) 7.6588920  0.0381288 40348.299   <2e-16 ***
kids        0.0094627  0.0070147     1.820   0.1773
age        -0.0002317  0.0009254     0.063   0.8023
disab      -0.0965115  0.0402883     5.739   0.0166 *
---
Signif. codes:  0 '***' 0.001 '**' 0.01 '*' 0.05 '.' 0.1 ' ' 1

Correlation structure = exchangeable
Estimated Scale Parameters:

            Estimate  Std.err
(Intercept)  0.05868 0.006549
  Link = identity

Estimated Correlation Parameters:
      Estimate Std.err
alpha   0.3692 0.04897
Number of clusters:    532  Maximum cluster size: 4
```

From the coefficients of the model, only the disab variable is significant.

```
gee_mod2 <- geeglm(hrs~kids + age + disab,
  data=workforce, id=id,
  corstr="independence", family = "gaussian")
```

Generalized Estimating Equations (GEE)

In the second model, gee_mod2, we specify the correlation matrix as "independence".

```
summary(gee_mod2)

Call:
geeglm(formula = hrs ~ kids + age + disab, family = "gaussian",
    data = workforce, id = id, corstr = "independence")

 Coefficients:
            Estimate    Std.err      wald Pr(>|w|)
(Intercept) 7.642883  0.038116 40207.83    <2e-16 ***
kids        0.010308  0.007243      2.03    0.1547
age         0.000208  0.000916      0.05    0.8206
disab      -0.119310  0.038938      9.39    0.0022 **
---
Signif. codes:  0 '***' 0.001 '**' 0.01 '*' 0.05 '.' 0.1 ' ' 1

Correlation structure = independence
Estimated Scale Parameters:

            Estimate Std.err
(Intercept)   0.0586 0.00655
Number of clusters:   532  Maximum cluster size: 4
```

Looking at the coefficients, again only the disab variable is significant in the model. This confirms that disab is the only variable significantly influencing the number of working hours within this sample.

```
QIC(gee_mod1); QIC(gee_mod2)

    QIC    QICu Quasi Lik    CIC   params    QICC
 136.39  132.87    -62.44   5.76     4.00  136.42
    QIC    QICu Quasi Lik    CIC   params    QICC
 143.33  132.79    -62.40   9.27     4.00  143.35
```

If we want to choose which model is the best one, we can use the *QIC*() function from the MESS package to get the QIC values of the models and choose the model with the lowest QIC (Quasi Information Criterion) as the best model. The gee_mod1 model has a QIC value of 136.39, lower than the QIC value of gee_mod2, therefore, it is the better model out of the two. The coefficients of the model are interpreted just like a regular linear model, where people who are disabled, on population average, work 0.0965 hours less than without disability.

Summary

Now that you have learned how to perform repeated measures ANOVA, build linear mixed models and GEE models, it's time to try it for yourself by taking some longitudinal data and analyzing it yourself by choosing the best method for asking questions about the dataset. If your questions require you to build a linear mixed model, maybe you can try to include more random and fixed effects than are shown in this chapter. If your questions require you to build a GEE model, you can try to build a model with a binomial outcome (like a logistic model) and try out different correlation structures to find your best possible model!

Further reading

If you want to learn more about longitudinal analysis in detail, there are many other resources that will explain all the methods I included in this chapter in detail (Diggle *et al*, 2002; Fitzmaurice *et al*, 2004).

Chapter 5 – Multivariate Analysis

Sections include

Discriminant Analysis
  Linear Discriminant Analysis (LDA)
  Quadratic Discriminant Analysis (QDA)
  Regularized Discriminant Analysis (RDA)

Canonical Correlation Analysis

Multidimensional Scaling (MDS)
  Metric MDS
  Nonmetric MDS

Introduction

When you read or hear the expression 'multivariate analysis', the first statistical concept that comes to mind may be building a regression model with one outcome variable and multiple independent variables that influence the outcome. However, if your knowledge of statistical concepts is a little deeper (or after you've read this book, a little bit broader), you may know another branch of statistics also referred to as **multivariate statistical analysis** (or **multi-outcome analysis**) that includes more statistical methods than just multivariate regression. This chapter includes three methods under the multivariate analysis umbrella. You may have heard of them if you have studied machine learning: **Discriminant analysis**, **Canonical correlation analysis**, and **Multidimensional scaling (MDS)** (Everitt and Hothorn, 2011).

Discriminant Analysis

Discriminant analysis is a great method to use when we need to categorize the individuals in a dataset into different groups based on the differences between their characteristics. The method is similar to a regression model in the sense that it calculates the weights (numerical impact) of the characteristics in order to predict which individual belongs in which group.

In this section, we will go through three different types of discriminant analysis method: **Linear**, **Quadratic**, and

Regularized. The methods are all run quite similarly in R, and each has its own pros and cons. Since discriminant analysis is a method for classifying subjects into groups, the best method is determined by the accuracy of the results.

As with all statistical methods, the best way to learn these different discriminant analysis methods is to use an example. We will use the dataset 'Skulls', which contains four different skull measurements (maximal breadth 'mb', basibregmatic 'height' bh, basialiveolar length 'bl', and nasal height 'nh') from ancient Egyptian skulls from different time periods (4000 BC, 3300 BC, 1850 BC, 200 BC, and 150 AD) found during archaeological digs. We can use discriminant analysis methods to classify the skulls into the time periods they were from by using the four skull measurements.

```
install.packages(c("MASS", "biotools",
    "mvnormtest", "Hmisc", "klaR"))
library(MASS); library(biotools);
    library(mvnormtest); library(Hmisc);
    library(klaR)
Skulls <- read.csv("Skulls.csv")
```

We can start by installing the R packages we need for the analyses: MASS, biotools, mvnormtest, Hmisc and klaR. We can install them using the install.packages() function and open them in the R environment using the library() function. We then open the dataset 'Skulls'. In this dataset, the epoch (time periods) is the outcome variable that we are interested in while the other four variables (skull measurements) are the independent variables.

Assumptions

Before we start looking at the discriminant analysis methods, let's test the assumptions for the Skulls dataset to see
if it is suitable for them. In order to use linear discriminant
analysis, the first method shown here, the dataset used must
meet four assumptions (Klecka, 1980):
1. Groups must be mutually exclusive (one individual cannot
belong to more than one group) and have equal sample sizes
2. Groups must have equal variance-covariance matrices
3. Independent variables should be normally distributed
4. Independent variables are not highly correlated

```
# Multivariate normality Test
shapiro.test(Skulls$mb);shapiro.test(Skulls$bh)
shapiro.test(Skulls$bl);shapiro.test(Skulls$nh)

        Shapiro-wilk normality test
data:  Skulls$mb
W = 0.99133, p-value = 0.4925

        Shapiro-wilk normality test
data:  Skulls$bh
W = 0.98643, p-value = 0.1503

        Shapiro-wilk normality test
data:  Skulls$bl
W = 0.99397, p-value = 0.7887

        Shapiro-wilk normality test
data:  Skulls$nh
W = 0.98405, p-value = 0.0804
```

Now that the Skulls dataset is open, we can start by testing
the normality of the dataset using the Shapiro-Wilk test,

which can be run using the shapiro.test() function from the mvnormtest package. Each output of the test for all four independent variables — mb, bh, bl, and nh — has a p value above 0.05, which means that we cannot reject the null hypothesis that the variables are normally distributed. This means that the normality assumption assumption is met.

```
# BoxM Test
boxM(Skulls[,2:5], Skulls[, 1])
```

Having met the normality assumption, we can see if the dataset satisfies the equality of variance-covariance matrix assumption. This is to see if the variance-covariance matrix (in other words, the errors of the model built from the dataset) are the same between the different groups of individuals (in this case, the different time periods). This can be easily tested using the Box's M Test using the boxM() function from the biotools R package. All we need to do is to place the columns 2 to 5 (of the datasets that contains the independent variables into the first argument of the boxM() function and the group variable column (column 1) into the second argument.

```
        Box's M-test for Homogeneity of Covariance Matrices
data: Skulls[, 2:5]
chi-Sq (approx.) = 45.667, df = 40, p-value = 0.2483
```

From the output of this test, we can see that the p value is higher than 0.05, which means that we cannot reject the null hypothesis that the variance-covariance matrices of the different time periods are the same. Put another way, the homogeneity assumption of the variance-covariance matrices has been met.

```
rcorr(as.matrix(Skulls[,2:5]))
```

Last but not least, we need to make sure that the independent variables are not highly correlated. For this assumption, we can use the rcorr() function from the Hmisc package. All we have to do is enter the columns 2 to 5 from the Skulls dataset Skulls[,2:5].

```
        mb    bh    bl    nh
mb   1.00 -0.06 -0.16  0.18
bh  -0.06  1.00  0.26  0.15
bl  -0.16  0.26  1.00 -0.01
nh   0.18  0.15 -0.01  1.00

n= 150


P
       mb     bh     bl     nh
mb        0.4517 0.0551 0.0254
bh 0.4517        0.0011 0.0731
bl 0.0551 0.0011        0.9382
nh 0.0254 0.0731 0.9382
```

The output of this test gives us all possible correlations of the four independent variables. We can see that generally the combinations are not statistically significant, except for the pairs mb and nh, and bl and bh. Although they are deemed to be statistically significantly correlated, the correlations are 0.18 and 0.26 respectively, which are not of too high a degree. So it may still be worthwhile to use them. Now that we have tested for the assumptions for discriminant analysis for this dataset, it is time to apply the data to linear discriminant analysis.

Linear Discriminant Analysis (LDA)

Linear discriminant analysis (LDA) is also known as **normal discriminant analysis** or **discriminant function analysis**. The linear discriminant analysis model looks quite like a linear regression model:

$DiscriminantScore =$
  $\alpha + \beta1 * IndepVar1 + \beta2 * IndepVar2 + \beta3 * IndepVar3$

where the weights are assigned to each independent variable so that the discriminant scores of each individual in the dataset are maximized to ensure highest likelihood of sorting the individuals into the correct groups (James *et al*, 2021).

```
lda_model <- lda(epoch~., data=Skulls)
lda_model
```

In the R environment, linear discriminant analysis can be performed using the lda() function from the MASS package. The first entry into this function is epoch~., which means in this case the outcome variable is epoch (the time periods where each individual will be classified) and the '.' after the ~ means the independent variables are the variables in the dataset that are not the epoch variable. We assign the data as the Skulls dataset and assign this function the name lda_model. When we run lda_model by itself, we can take a look at the details.

```
Call:
lda(epoch ~ ., data = Skulls)

Prior probabilities of groups:
c1850BC  c200BC c3300BC c4000BC  cAD150
   0.2     0.2     0.2     0.2     0.2

Group means:
                mb        bh        bl        nh
c1850BC  134.4667 133.8000 96.03333 50.56667
c200BC   135.5000 132.3000 94.53333 51.96667
c3300BC  132.3667 132.7000 99.06667 50.23333
c4000BC  131.3667 133.6000 99.16667 50.53333
cAD150   136.1667 130.3333 93.50000 51.36667

Coefficients of linear discriminants:
           LD1          LD2          LD3          LD4
mb  0.12667629 -0.03873784 -0.09276835 -0.1488398644
bh -0.03703209 -0.21009773  0.02456846  0.0004200843
bl -0.14512512  0.06811443 -0.01474860 -0.1325007670
nh  0.08285128  0.07729281  0.29458931 -0.0668588797

Proportion of trace:
   LD1    LD2    LD3    LD4
0.8823 0.0809 0.0326 0.0042
```

When we use the lda() function for linear discriminant analysis, the same number of linear discriminant analysis models are built as the independent variables that are under consideration (in our case, four models are built). The model that is used to classify individuals is the model that explains the highest percentage of variance (differences) between the individuals. In the outcomes of lda_model, the prior probabilities of groups show the division of the proportion of individuals in each time period. The group means represent the means of each of the four independent variables at each time period in the original dataset. The coefficients of the linear discriminants represent the coefficients of the independent variables on the subject group classification for each of the four models built. The 'Proportion of trace' is the percentage of variance explained by each of the four models built. In our case, the best linear discriminant model built with the Skulls dataset (LD1) explains 88.23% of variance of the dataset and its coefficients for the independent variables (mb, bh, bl, nh) are approximately 0.13, -0.04, -0.15, and 0.083 respectively.

```
predict_lda = predict(lda_model)$class
original_lda = Skulls$epoch
```

Now that we have built our linear discriminant model, we can use this model to predict the time period group membership of each individual in the dataset. To do this, we can enter the function name lda_model into the predict() function and add $class to the end of the predict() function to get the class portion of the prediction (which contains the predicted values). We can assign the predicted time period values the name predict_lda. We can also get the actual values of the time periods using Skulls$epoch and assign these values the name original_lda.

```
out_lda = data.frame(original_lda, predict_lda)
```

We can put these two columns of original time periods and predicted time periods together by using the data.frame() function and calling the dataframe out_lda. This dataframe is used to determine the percentage of people whose time periods were classified correctly using the model we built.

```
Correct_lda <-
    as.character(out_lda$original_lda) ==
    as.character(out_lda$predict_lda)
```

To get the percentage of the correctly predicted individuals, we can take the original_lda and predict_lda columns of the out_lda dataframe and then convert them both into characters using the as.character() function. Then we can compare these two character columns using the '==' sign. What this returns is a list filled with binary values TRUE and FALSE, which are the results of comparing the values in the original_lda with their counterparts in predict_lda. We can assign this list of TRUE/FALSE values the name correct_lda.

```
sum(Correct_lda)/150
```

We can use the function sum() to add up all the TRUEs (where the predicted time period is the same as the original time periods) and divide this number by the total number of individuals in the dataset, 150.

```
[1] 0.34
```

From the output, we can see that 34% of the predictions were correct.

```
# Significance of the classification
chisq.test(out_lda$original_lda,
    out_lda$predict_lda)

        Pearson's Chi-squared test
data:  out_lda$original_lda and out_lda$predict_lda
X-squared = 48.107, df = 16, p-value = 4.57e-05
```

We can also test for the statistical significance of the percentage of accurate predictions by using the chi-square test on the original time periods and the predicted time periods by putting the original_lda and predict_lda columns into the chisq.test() function. From the output of the chi-square test, the *p* value is below 0.05, which means the discriminant analysis model with the coefficients listed in LD1, significantly predicts the time periods from which the skulls come from better than a model where all four coefficients are 0.

Quadratic Discriminant Analysis (QDA)

Now that we have gone through discriminant analysis using the linear discriminant analysis method, you will be glad to learn that there are other methods that can accommodate datasets that do not perfectly meet the assumptions that are listed for the linear discriminant analysis method (James *et al*, 2021). Quadratic discriminant analysis is a method that is useful when you know that the variance-covariance matrices between the classification groups are different.

```
qda_model <- qda(epoch~., data=Skulls)
qda_model
```

In R, a QDA model only differs from LDA by using the qda() function (from the MASS package). Here, we name the model qda_model. Running qda_model by itself, we can get the group probabilities of the original dataset and the group means of the independent variables in the original dataset:

```
Call:
qda(epoch ~ ., data = Skulls)

Prior probabilities of groups:
c1850BC   c200BC c3300BC c4000BC   cAD150
    0.2      0.2     0.2     0.2      0.2

Group means:
              mb       bh       bl       nh
c1850BC 134.4667 133.8000 96.03333 50.56667
c200BC  135.5000 132.3000 94.53333 51.96667
c3300BC 132.3667 132.7000 99.06667 50.23333
c4000BC 131.3667 133.6000 99.16667 50.53333
cAD150  136.1667 130.3333 93.50000 51.36667
```

```
predict_qda = predict(qda_model)$class
original_qda = Skulls$epoch
out_qda = data.frame(original_qda, predict_qda)
```

Now that the model is built, we can use the model to classify the individual skulls in the dataset into the time periods and put the original time periods and the predicted time periods into a dataframe together using the predict() and data.frame() functions. We get a dataframe called out_qda that contains the columns original_qda and predict_qda for the original and predicted time periods.

```
Correct_qda <-
    as.character(out_qda$original_qda) ==
    as.character(out_qda$predict_qda)

sum(Correct_qda)/150

[1] 0.36
```

After that, we can compare the predicted classification with the original classification in the same manner as LDA and name the list of TRUE/FALSE values Correct_qda and calculate the percentage of correctly predicted time periods, which is 36%.

```
chisq.test(out_qda$original_qda,
    out_qda$predict_qda)

        Pearson's Chi-squared test

data:  out_qda$original_qda and out_qda$predict_qda
X-squared = 41.327, df = 16, p-value = 0.0004968
```

Using the chi-square test to check significance of the model's ability to correctly classify the skulls, we can see the p value (as in LDA) is lower than 0.05, therefore, the model predicted the time periods of the skulls significantly.

Since the QDA model predicted 36% of the time periods correctly, higher than the 34% in the LDA and both models are significant, QDA is the better model compared to LDA model for the Skulls dataset for predicting the time periods.

Regularized Discriminant Analysis (RDA)

Like QDA, regularized discriminant analysis is another alternative to LDA. It allows the built model to be more robust against correlations in the independent variables. Another additional thing we need to pay attention to when using linear or quadratic discriminant analyses is that the number of members in each of the groups cannot be lower than the number of independent predictor variables. Regularized discriminant analysis, however, can bypass this requirement when we have a big dataset where the number of independent variables may be higher than the number of individuals (Friedman, 1989). Also like QDA, the regularized discriminant analysis model's building process is almost identical to LDA, using the rda() function from the klaR package.

```
rda_model <- rda(epoch~., data=Skulls)
rda_model
```

We can name the RDA model rda_model and when we run this model by itself, we get a new output called regularization parameters. Regularization parameters (gamma and lambda) both take on values between 0 to 1. When gamma and lambda are both equal to 0, we have a QDA model; when lambda is equal to 1 while gamma is equal to 0, we have an LDA model. In RDA models, the regularization parameters are what makes the model unique.

```
Call:
rda(formula = epoch ~ ., data = Skulls)

Regularization parameters:
    gamma     lambda
0.5406939 0.2390350

Prior probabilities of groups:
c1850BC   c200BC c3300BC c4000BC   cAD150
    0.2      0.2      0.2      0.2      0.2

Misclassification rate:
      apparent: 61.333 %
cross-validated: 71.333 %
```

The output also contains the misclassification rate (the percentage of the individuals that will be misclassified by the model. For our model, 61.333% of the individual skulls in the dataset will be classified incorrectly by the model (do note some variability in results each time this is carried out).

```
predict_rda = predict(rda_model)$class
original_rda = Skulls$epoch
out_rda = data.frame(original_rda, predict_rda)

Correct_rda <-
    as.character(out_rda$original_rda) ==
    as.character(out_rda$predict_rda)
sum(Correct_rda)/150
```

```
[1] 0.3866667
```

We can double check this number by creating predicted values and calculate the percentage of the correctly predicted time periods for skulls using the same method as LDA and QDA, and the correctly predicted percentage is 38.66667%, which adds up to 100% with the misclassified. (Again, notice that the result will differ somewhat each time QDA is used.)

```
chisq.test(out_rda$original_rda,
    out_rda$predict_rda)
```

```
        Pearson's Chi-squared test
data:  out_rda$original_rda and out_rda$predict_rda
X-squared = 48.87, df = 16, p-value = 0.00003464
```

We can also test the significance of the predictions of this model using the chi-square test. The p value this time, as with LDA and QDA, is also below 0.05 (but comes out very slightly different each time RDA is run), which means the model's ability to predict values is significant. This model can also be seen as a better option than the LDA model due to the higher correct prediction percentage.

Other methods

LDA, QDA, and RDA are not the only discriminant analysis methods. Others include **mixture discriminant analysis (MDA)** and **flexible discriminant analysis (FDA)**. Both of them can be performed using a similar method as the methods previously described, using the mda() and fda() functions respectively from the *mda* package. In normal circumstances where discriminant analysis is used to classify cases, multiple models are built using different methods; the final model chosen for the final prediction is the model with the highest correct prediction percentage. Try using your own dataset, using all the methods explained here to build your own models and see how well they can perform!

Canonical Correlation Analysis

Canonical correlation analysis is another multivariate statistical analysis method where there are multiple dependent variables of interest. Unlike discriminant analysis, it is not a method used to classify study subjects. It is used to identify and measure the associations among two sets of variables in a dataset (Hotelling, 1936). Canonical correlation is especially useful in situations where there are multiple intercorrelated outcome variables. Its purpose is to look at whether or not the independent variables can be used to significantly predict each of the outcome variables. In addition, we are also interested in how many dimensions are significant for the dataset. In a sense, CCA is very similar to multiple regression with an extra step.

Canonical correlation's assumption tests are similar to other multivariate statistics as well as related to the correlation among dependent and independent variables. There are two important assumptions:
1. Normality — The sets of independent and dependent variables must have multivariate normal distributions.
2. Linearity — correlation between the independent and dependent variables cannot be zero.

In order to best understand canonical correlation analysis, we can go through an example. The dataset *sales* (Pennsylvania State University, 2022) is a dataset which contains seven variables representing information on the sales performance of employees ('SalesGrowth', 'SalesProfitability', and 'NewAccountSales') and employees' intelligence test scores ('Creativity', 'MechanicalReasoning', 'AbstractReasoning', and 'Mathematics'). In our example, the sales performance variables are considered as the dependent variables while the intelligence measurements of the employees are considered as independent variables. Our interest is to see whether dependent variables and the independent variables are correlated.

```
install.packages(c("CCA", "yacca", "MVN"))
library(CCA); library(yacca); library(MVN)
sales = read.csv("sales.csv")
```

In order to perform canonical correlation analysis on this dataset, we need two R packages *yacca* and *CCA*, with the *MVN* package to test for multivariate normality. (Occasionally, installations of the MVN package are hindered by a failure to install the *car* package, something to do with locking. See the footnote for the solution. *) To read the dataset into the R environment, we can use the read.csv() function, assigning the object with the name 'sales'.

```
salesperformance <- sales[, c("SalesGrowth",
   "SalesProfitability", "NewAccountSales")]
testperformance <- sales[, c("Creativity",
   "MechanicalReasoning", "AbstractReasoning",
   "Mathematics")]
```

We can separate out the variables for personal attributes and the sales performance measures, turning them into separate

---

* `install.packages("car", dependencies = TRUE, INSTALL_opts =` '`-no-lock`'), then `install.packages("MVN")`.

datasets called 'salesperformance' and 'testperformance'. To do this, we enter the names of the variables for each subset (as strings) into a vector, then placing the vector to the right of the comma within the square brackets.

```
mvn(salesperformance)
```

```
$multivariateNormality
          Test       HZ     p value MVN
1 Henze-Zirkler 1.28537 0.000833093  NO


$univariateNormality
            Test          Variable Statistic  p value Normality
1 Anderson-Darling     SalesGrowth    1.0433   0.0088    NO
2 Anderson-Darling SalesProfitability 0.5266   0.1710    YES
3 Anderson-Darling  NewAccountSales   0.3372   0.4916    YES


$Descriptives
...
```

```
mvn(testperformance)
```

```
$multivariateNormality
          Test        HZ    p value MVN
1 Henze-Zirkler 0.8075985 0.3168453 YES


$univariateNormality
            Test          Variable Statistic  p value Normality
1 Anderson-Darling      Creativity    0.6520   0.0838    YES
2 Anderson-Darling MechanicalReasoning 0.4681   0.2394    YES
3 Anderson-Darling AbstractReasoning   1.3037   0.0020    NO
4 Anderson-Darling     Mathematics    0.3760   0.3993    YES


$Descriptives
...
```

We can start by testing the normality of the salesperformance and testperformance variables by using the mvn() function. We can see from the univariate normality sections of the outputs that variables in both of the new datasets are not all normal, therefore, neither dataset can achieve the multivariate normal assumption. In situations like this, we can either remove the variables from the datasets or transform them so that they will become normal.

```
salesperformance <-
    sales[, c("SalesProfitability",
    "NewAccountSales")]
testperformance <-
    sales[, c("Creativity",
    "MechanicalReasoning", "Mathematics")]
```

In this example, we remove the non-normal variables and give the datasets salesperformance and testperformance the remaining 'normal' variables. We can test for the normality of the datasets again using the mvn() function and we can see that both datasets now meet the normality assumption, with each of the variables having a normal distribution and each dataset having achieved multivariate normality (as stated in the multivariate normal section of the output).

```
mvn(salesperformance)
```

```
$multivariateNormality
          Test        HZ   p value MVN
1 Henze-Zirkler 0.7218473 0.1309474 YES

$univariateNormality
          Test           Variable Statistic  p value Normality
1 Anderson-Darling SalesProfitability   0.5266   0.1710    YES
2 Anderson-Darling  NewAccountSales     0.3372   0.4916    YES

$Descriptives
...
```

mvn(testperformance)$multivariateNormality

```
          Test        HZ   p value MVN
1 Henze-Zirkler 0.7236161 0.3032234 YES
```

mvn(testperformance)$univariateNormality

```
          Test            Variable Statistic  p value Normality
1 Anderson-Darling      Creativity    0.6520   0.0838    YES
2 Anderson-Darling MechanicalReasoning 0.4681   0.2394    YES
3 Anderson-Darling      Mathematics   0.3760   0.3993    YES
```

mvn(testperformance)$Descriptives

```
                    n  Mean   Std.Dev Median Min Max 25th 75th     Skew    Kurtosis
Creativity          50 11.22  3.950149   10.0   1  18 8.25   14  0.05576043 -0.4944719
MechanicalReasoning 50 14.18  3.384780   15.0   5  20 12.00   17 -0.31586390 -0.4007025
Mathematics         50 29.76 10.537707   31.5   9  51 21.50   37 -0.10100835 -0.8971707
```

Next, we can compute the correlation of the dependent variables and independent variables among themselves as well as among each other. To do this, we can use the matcor() function from the *CCA* package. The outputs $Xcor and $Ycor contain the correlations between the variables within each dataset; \$XYcor includes correlations across both datasets. As we can see from the output, the correlations between different variables are all above 0 and below 1 (as all correlation values should be), which means the linearity assumption stands.

```
matcor(salesperformance, testperformance)
```

```
$xcor
                   SalesProfitability NewAccountSales
SalesProfitability          1.0000000       0.8425232
NewAccountSales             0.8425232       1.0000000

$ycor
                   Creativity MechanicalReasoning Mathematics
Creativity          1.0000000           0.5907360   0.4126395
MechanicalReasoning 0.5907360           1.0000000   0.5745533
Mathematics         0.4126395           0.5745533   1.0000000

$XYcor
                   SalesProfitability NewAccountSales Creativity MechanicalReasoning Mathematics
SalesProfitability          1.0000000       0.8425232  0.5415080           0.7459097   0.9442960
NewAccountSales             0.8425232       1.0000000  0.7003630           0.6374712   0.8525682
Creativity                  0.5415080       0.7003630  1.0000000           0.5907360   0.4126395
MechanicalReasoning         0.7459097       0.6374712  0.5907360           1.0000000   0.5745533
Mathematics                 0.9442960       0.8525682  0.4126395           0.5745533   1.0000000
```

Before we begin the canonical correlation analysis, let's gain a rough understanding of the idea of canonical correlation analysis. We have two groups of variables (sales performance and intelligence test performance). We can write the variables in each group into linear combinations:

$Salesperformance_1 =$
$\quad \beta_{1SP} SalesProfitability \ + \ \beta_{1NAS} NewAccountSales$
$Salesperformance_2 =$
$\quad \beta_{2SP} SalesProfitability \ + \ \beta_{2NAS} NewAccountSales$
$Testperformance_1 =$
$\quad \beta_{1C} Creativity \ + \ \beta_{1MR} MechanicalReasoning$
$\quad + \ \beta_{1M} Mathematics$
$Testperformance_2 =$
$\quad \beta_{2C} Creativity \ + \ \beta_{2MR} MechanicalReasoning$
$\quad + \ \beta_{2M} Mathematics$

$Testperformance_3 =$
    $\beta_{3C}Creativity + \beta_{3MR}MechanicalReasoning$
    $+ \beta_{3M}Mathematics$

Salesperformance$_1$, Salesperformance$_2$, etc., are canonical variates; each β is a unique coefficient for the variable in the linear combination; and the number of linear combinations is the same as the number of variables in each group. The linear combinations with the same numbers are called *canonical variate pairs*; in this example, there are two canonical variate pairs — Salesperformance$_1$, Testperformance$_1$; Salesperformance$_2$, Testperformance$_2$ — because the number of canonical variate pairs depends on how many variables are in the smaller group of variables.

The purpose of canonical correlation analysis is to find linear combinations that maximize the correlations between the members of each canonical variate pair. To do this, we can use the cc() function from the CCA package and enter the two sets of variables, salesperformance and testperformance, into its arguments. We give the results of this canonical correlation analysis the name cc_sales. If we run cc_sales by itself, the outputs are the results of the analysis.

```
cc_sales = cc(salesperformance,testperformance)
cc_sales
```

```
$cor
[1] 0.9882891 0.5340325
```

Under $cor, we have the canonical correlations of the two canonical variate pairs. The canonical correlation for the first canonical variate pair —

Salesperformance$_1$, Testperformance$_1$ — is 0.988, the higher correlation of the members of each canonical variate pair.

```
$xcoef
                      [,1]        [,2]
SalesProfitability  -0.07091005 -0.1691005
NewAccountSales     -0.06734402  0.3881685

$ycoef
                           [,1]         [,2]
Creativity              -0.04693576  0.30876752
MechanicalReasoning     -0.05713992 -0.25476069
Mathematics             -0.07326936 -0.01090848
```

$xcoef and $ycoef are the raw coefficients for the salesperformance and testperformance variables; they can be written as canonical variate functions shown below. The raw coefficients can be interpreted the same way as in a linear regression where, for example, if Salesprofitability increases by 1, Salesperformance$_1$ will decrease by -0.07091005.

$$Salesperformance_1 =$$
$$- 0.07091005(SalesProfitability)$$
$$- 0.06734402(NewAccountSales)$$
$$Testperformance_1 =$$
$$- 0.04693576(Creativity)$$
$$- 0.05713992(MechanicalReasoning)$$
$$- 0.07326936(Mathematics)$$
$$Salesperformance_2 =$$
$$- 0.1691005(SalesProfitability)$$
$$+ 0.3881685(NewAccountSales) Testperformance_2 =$$
$$+ 0.30876752(Creativity)$$
$$- 0.25476069(MechanicalReasoning)$$
$$- 0.01090848(Mathematics)$$

Now that we have our canonical correlation for the canonical variates, we need to test the correlations to see whether they are statistically significant. We can do this by using the cca() function from the yacca package to rerun the canonical correlation analysis and we name the outcomes cca_sales. We can then perform an F test on the results using the F.test.cca() function from the yacca package.

```
cca_sales <-
    cca(salesperformance, testperformance)
F.test.cca(cca_sales)
```

```
        F Test for Canonical Correlations (Rao's F Approximation)

          Corr         F   Num df Den df              Pr(>F)
CV 1   0.98829 101.26832  6.00000     90 < 0.00000000000000022 ***
CV 2   0.53403       NaN  2.00000    NaN                   NaN
---
Signif. codes:  0 '***' 0.001 '**' 0.01 '*' 0.05 '.' 0.1 ' ' 1
```

From the results, we can see that only the canonical correlation from the first canonical variate pair is significant, with a p value below 0.05. This means we can interpret the coefficients for the first canonical variate, as statistically significant, and we can claim that the two sets of variables for sales performance and intelligence test performance are correlated.

Multidimensional Scaling

Multidimensional Scaling (MDS) is an area of multivariate analysis that helps to visualize the distances (the level of similarity or dissimilarity) between subjects (Borg and Groenen, 2005). A dataset used in this kind of analysis comprises a matrix, a square shaped dataset where there is one column and one row of data for each subject in the dataset. In this section, we are going to go through two types of MDS: **Metric MDS** (also known as Principal Coordinates Analysis) is used when the distances are measured in interval data and there is only one distance for each pair of subjects (that is, there is no measurement error); **Nonmetric MDS** is when the distances are ordinal and there is only one distance for each pair of subjects. To put it in plain terms, MDS takes the dissimilarities (distances) of the individuals in a dataset and a chosen number of dimensions (as in a plot) and tries to place each of the individuals in the plot while preserving the distances between the individuals as much as possible. We will be able to see this much more clearly once we go through an example.

Metric MDS

Below are the distances in miles between the largest cities on each continent (ok, research station in the case of Antarctica). We can read the dataset into the R environment using the read.csv() function (specify header=FALSE so the first line of the dataset is not used for column names) and call the dataset 'LargestPops'. We change the columns and the

rows so that they have the same names, using the colnames()
and rownames() functions.

```
LargestPops <-
   read.csv("LargestPops.csv", header=FALSE)
varnames <- c("Auckland", "Buenos Aires",
   "Istanbul", "Kinshasa", "McMurdo Station",
   "Mexico City", "Shanghai")
rownames(LargestPops) <- varnames
colnames(LargestPops) <- varnames
LargestPops # shows the dataset
```

	Auckland	Buenos Aires	Istanbul	Kinshasa	McMurdo Station	Mexico City	Shanghai
Auckland	0	6448	10588	9306	2846	6803	5815
Buenos Aires	6448	0	7600	5125	4458	4581	12197
Istanbul	10588	7600	0	3237	9618	7112	4974
Kinshasa	9306	5125	3237	0	6659	7916	7333
McMurdo Station	2846	4458	9618	6659	0	7581	7744
Mexico City	6803	4581	7112	7916	7581	0	8032
Shanghai	5815	12197	4974	7333	7744	8032	0

The dataset is a matrix of the distances between each of the
seven locations. Notice that the matrix's diagonal contains
zeroes. The following code computes the classical MDS:

```
install.packages(c("MASS", "psy", "ggplot2"))
library(MASS); library(psy); library(ggplot2)
dataframe <- LargestPops
# subsequent lines reusable
dist.data <-
    dist(scale(t(dataframe), center = TRUE,
    scale = TRUE), method = 'euclidean')
mMDS <- cmdscale(dist.data, eig = TRUE, k = 2)
mMDS
```

To compute the classical MDS,   we first calculate the
distance matrix: the t() function transposes the data, the
scale() function includes centering of rows and columns,
and the distance is the default Euclidean. The cmdscale()
function (from the MASS package) takes the transformed
data in its first argument, eig = TRUE, and k = 2. The eig =
TRUE expression allows the function to return eigenvalues,
which are values used to compute the coordinates of each
location on a plot. The k = 2 means that we choose two
dimensions. We name this metric ('classical') MDS process
as 'mMDS', running just the name to see the output:

```
$points
                      [,1]       [,2]
Auckland         2.0645174 -1.3767008
Buenos Aires     1.0954458  1.9873999
Istanbul        -2.6171134  0.2116080
Kinshasa        -1.6483012  0.9843447
McMurdo Station  1.8869041 -0.3338391
Mexico City      0.4816634  0.9233382
Shanghai        -1.2631160 -2.3961510

$eig
[1] 2.041628e+01 1.356432e+01 7.200681e+00 4.422945e-01 3.067820e-01
[6] 6.964394e-02 1.352705e-15


$GOF
[1] 0.8090619 0.8090619
```

The most important outputs are $points, $eig, and $GOF.
The first of these shows the visual coordinates, representing
the seven locations' dissimilarities. The second, $eig, shows
the eigen values, which compute the goodness of fit. $GOF,

the goodness of fit, contains two values, between 0 to 1, and should be as close as possible to 1 in order for the MDS to be of good fit. In our example, the values are both 0.8090619, trending towards a good fit, but would be better if closer to 1. (We could opt for k = 3, which would produce \$GOF values of over 0.9, but we will consider this in the next section.)

```
mMDS.value <- mMDS$points
# using the visual coordinates
mMDS.data <-
    data.frame(Sample1 = rownames(mMDS.value),
    Dim1 = mMDS.value[,1], Dim2 = mMDS.value[,2])
ggplot(data = mMDS.data,
    aes(Dim1, Dim2, label= Sample1)) +
    geom_text() + theme_bw()
```

The first piece of code utilizes the visual coordinates, the second creates a data frame, and ggplot() from the ggplot2 package creates the plot.

While the algorithm does not have a compass, the relationships created by distance make some sort of sense.

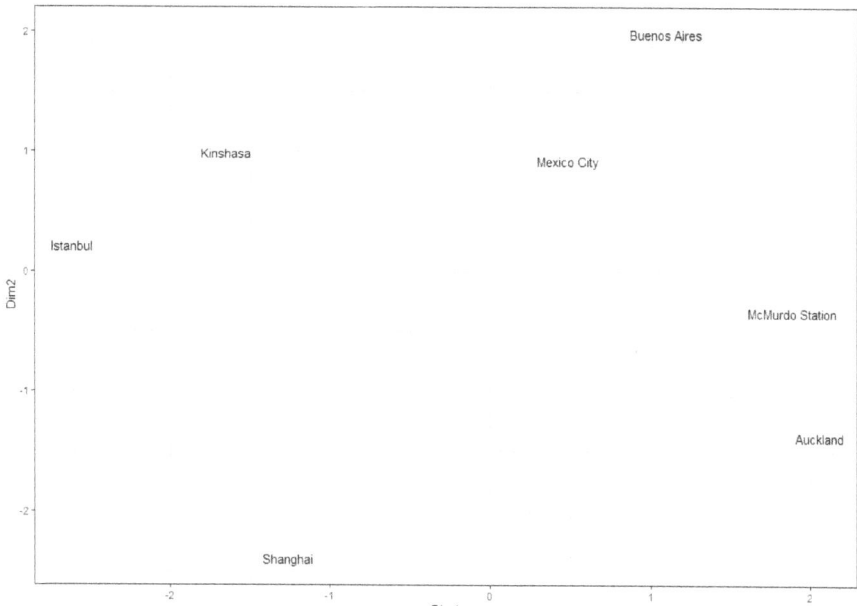

To check to see if 2 dimensions is the correct number of dimensions, we can use the scree.plot() function from the psy package, placing the LargestPops dataset within the function:

```
scree.plot(LargestPops, title = "Scree Plot",
    type="R")
```

Scree Plot

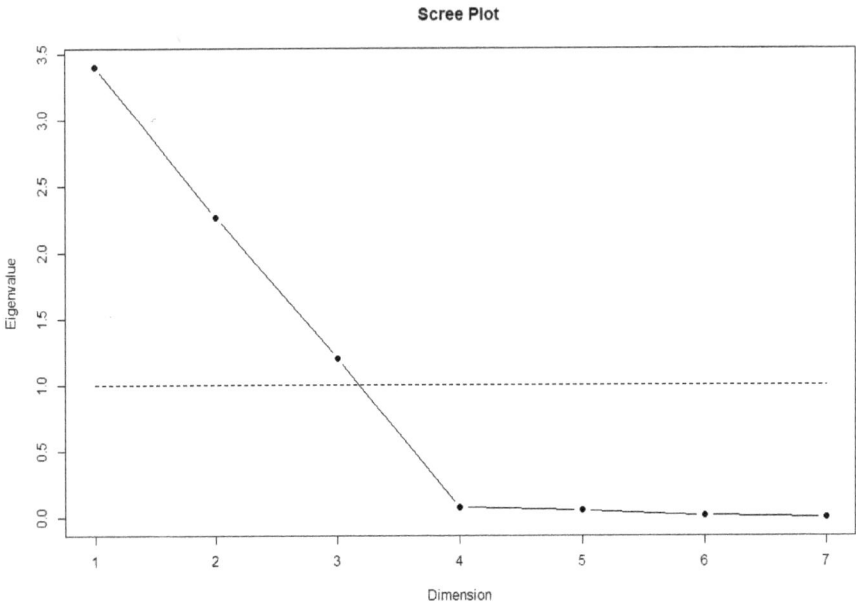

On the scree plot, we can see that 4 points are above the scree (foothill rubble), which indicates 4 dimensions as being optimal. The dotted line represents the Kaiser criterion, a threshold of 1 eigenvalue or more; this indicates 3 dimensions. I prefer the scree test for larger datasets, but Kaiser for small ones.

To recode, try something like this:

```
Rerun_mMDS <-
    cmdscale(dist.data, eig = TRUE, k = 3)
mMDS <- Rerun_mMDS
mMDS # follow on with mMDS.value and so on
```

While the k = 3 option (three dimensions) has a higher goodness of fit than k = 2, at 0.9805067, the chart shows no appreciable difference. However, as MDS visualization is quite a subjective matter, it might be worth trying a higher *k* value with larger and/or more complex datasets.

Nonmetric MDS

For Nonmetric MDS, the difference is in the dataset, in the sense that the typical matrix contains ordinal data rather than interval data. As an example, we can use the dataset NEO PI-R (NEO Personality Inventory – Revised; Costa and McCrae, 1992). This is a matrix comprising the correlations of 30 personality variables; there are 6 variables under each of the 5 broad personality categories: Openness, Conscientiousness, Extraversion, Agreeableness, and Neuroticism.

```
#MDS Non-Metric Example
install.packages(c("MASS", "psy", "psych",
    "vegan", "ggplot2"))
library(MASS); library(psy); library(psych);
library(vegan); library(ggplot2)
NEO_PI_R <-
    read.csv("NEO_PI_R.csv", header=FALSE)
```

To perform Nonmetric MDS, we need to use the library function to open five R packages — MASS, psy, psych, vegan, and ggplot2 — using the install.packages() function as necessary. We read the dataset into the R environment using read.csv() and name the dataset NEO_PI_R.

```
varnames <- c(
  "N1Anxiety", "N2AngryHostility",
  "N3Depression", "N4Self-Consciousness",
  "N5Impulsiveness" , "N6Vulnerability",
  "E1Warmth" , "E2Gregariousness",
  "E3Assertiveness", "E4Activity",
  "E5Excitement-Seeking", "E6PositiveEmotions",
  "O1Fantasy", "O2Aesthetics", "O3Feelings",
  "O4Ideas", "O5Actions" , "O6Values",
  "A1Trust", "A2Straightforwardness",
  "A3Altruism", "A4Compliance" , "A5Modesty",
  "A6Tender-Mindedness",
  "C1Competence", "C2Order", "C3Dutifulness" ,
  "C4AchievementStriving", "C5Self-Discipline",
  "C6Deliberation")
rownames(NEO_PI_R) <- varnames
colnames(NEO_PI_R) <- varnames
```

We can then place the names of all 30 variables into a vector and call it varnames. Then we use the rownames() and colnames() functions to give names to the dataset's rows and columns.

```
head(NEO_PI_R) # severely truncated image
```

	N1Anxiety	N2AngryHostility	N3Depression	N4Self-Consciousness	N5Impulsiveness
N1Anxiety	1.00	0.47	0.64	0.54	0.34
N2AngryHostility	0.47	1.00	0.52	0.37	0.40
N3Depression	0.64	0.52	1.00	0.60	0.38
N4Self-Consciousness	0.54	0.37	0.60	1.00	0.31
N5Impulsiveness	0.34	0.40	0.38	0.31	1.00
N6Vulnerability	0.60	0.43	0.63	0.56	0.35

As you will notice from this snapshot of the database, the diagonal where each element is correlated with itself is comprised of 1's. This is to be expected from a correlational dataset.

```
dist.data <- cor2dist(NEO_PI_R)
head(dist.data) # truncated
```

```
                  N1Anxiety N2AngryHostility N3Depression N4Self-Consciousness N5Impulsiveness
N1Anxiety         0.0000000        1.0295630    0.8485281            0.9591663       1.148913
N2AngryHostility  1.0295630        0.0000000    0.9797959            1.1224972       1.095445
N3Depression      0.8485281        0.9797959    0.0000000            0.8944272       1.113553
N4Self-Consciousness 0.9591663     1.1224972    0.8944272            0.0000000       1.174734
N5Impulsiveness   1.1489125        1.0954451    1.1135529            1.1747340       0.000000
N6Vulnerability   0.8944272        1.0677078    0.8602325            0.9380832       1.140175
```

The correlational dataset has been changed into a distance (difference) dataset using the cor2dist() function from the psych package. The dataset's diagonals can now be seen to contain 0's.

```
nmMDS <- metaMDS(dist.data, k=2, try= 100 )
nmMDS
```

To perform Nonmetric MDS, we need to use the metaMDS() function from the vegan package. We run the new object, 'NMDS' by itself:

```
Call:
metaMDS(comm = dist.data, k = 2, try = 100)

global Multidimensional Scaling using monoMDS

Data:       dist.data
Distance: user supplied

Dimensions: 2
Stress:     0.1430375
Stress type 1, weak ties
Best solution was repeated 11 times in 20 tries
The best solution was from try 0 (metric scaling or null solution)
Scaling: centring, PC rotation
Species: scores missing
```

The 'Stress' output is a value of 0 or greater that reflects how well the chosen dimension fits the nonmetric MDS. Ideally, the stress value should be as close to 0 as possible to make the nonmetric MDS a good fit. In our case, however, the value of 0.143 could be bettered. Another dimension should be chosen. As an aid, let's view a scree plot.

```
scree.plot(dist.data, title ="Scree Plot",
    type ="R")
```

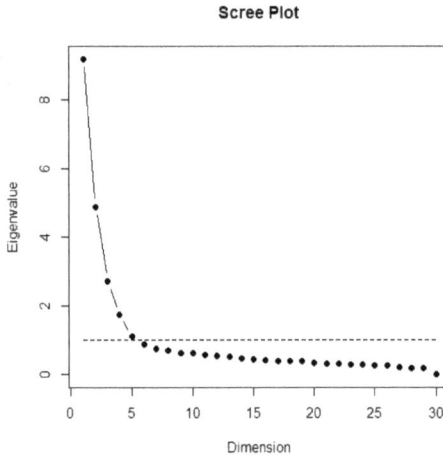

Scree Plot

If we follow the advice of the scree plot, and Kaiser's criterion, we run the metaMDS() function with k = 5:

```
Rerun_nmMDS <-
    metaMDS(dist.data, k=5, try= 100 )
Rerun_nmMDS
```

```
Call:
metaMDS(comm = dist.data, k = 5, try = 100)

global Multidimensional Scaling using monoMDS

Data:      dist.data
Distance: user supplied

Dimensions: 5
Stress:     0.03515643
Stress type 1, weak ties
Best solution was repeated 1 time in 20 tries
The best solution was from try 17 (random start)
Scaling: centring, PC rotation
Species: scores missing
```

The Stress value has lowered to 0.035, which shows if we have 5 dimensions it will be a better fit. However, we must note that the stress value will be decreased as long as we keep increasing the k value, and a stress value of 0 may be meaningless due to overfitting. This is also the case when we increase the k values in metric MDS, affecting the GOF values. Therefore, we should use the scree plot to make the final decision on how many dimensions to choose.

You may notice when running metaMDS() that the values in the last decimal places may be different from those shown. The 'random starts' (each called a 'try' in the code) create a non-deterministic algorithm!

Now let's have the visualization:

```
actualNonMetric <- Rerun_nmMDS
# assigning to actualNonMetric
# allows reusable code:

actualNonMetric.value <- actualNonMetric$points
  # uses visual coordinates
actualNonMetric.data <- data.frame(Sample2 =
  rownames(actualNonMetric.value),
  Dim1 = actualNonMetric.value[,1],
  Dim2 = actualNonMetric.value[,2])
ggplot(data = actualNonMetric.data,
  aes(Dim1, Dim2, label= Sample2)) +
  geom_text() +
  theme_bw()
```

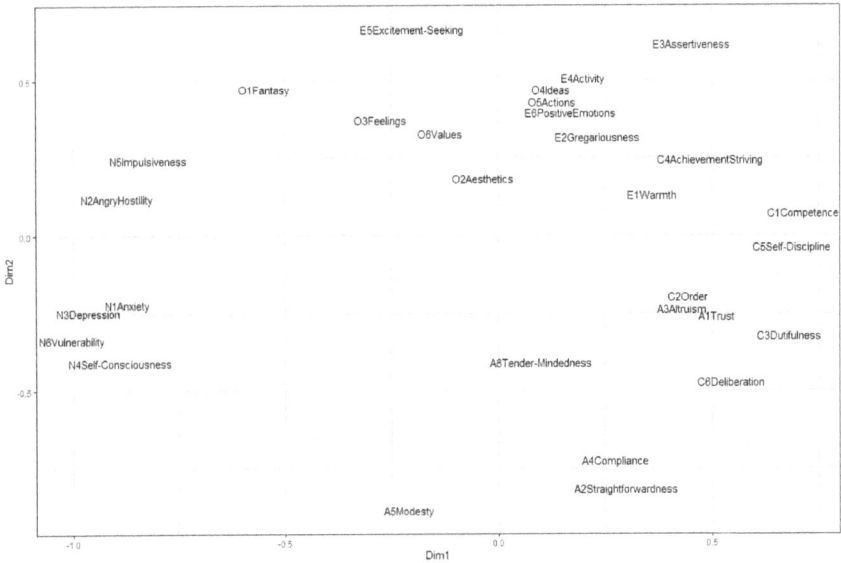

Summary

In this chapter, we have gone through discriminant analysis, canonical correlation analysis, and multidimensional scaling. These three are not the only topics in multivariate analysis: exploratory factor analysis, principal component analysis, MANOVA and MANCOVA are useful methods that are not included in this book, but appear in its sister book (Davis, 2022).

The methods from this chapter have been covered in greater depth by books and papers cited at the end of each section. If you only wish to use these methods without the need to understand them in great depth, then it is time for you to apply these methods to your own datasets and use the broad, application-level knowledge this chapter (hopefully) has given you.

Further reading

For discriminant analysis, Silva and Stam (1995) and Stevens (2002) are recommended reading. On canonical correlational analysis, Thompson (2000) and Stevens (2002) are useful. Stalans (1995) is useful for multidimensional scaling.

Chapter 6 – Miscellaneous methods

Sections include

Generalized Linear Models (GLM)

Poisson regression

Hierarchical Modeling

Power Analysis

Reliability
    Test-retest reliability
    Internal reliability
    Inter-rater reliability

Introduction

Now that you have gone through all of the previous chapters (or most of the book), you probably now have a rough idea of how some of the statistical methods you were wondering about worked and how to apply them to a dataset or two. This chapter covers a few miscellaneous topics not covered in the previous chapters: generalized linear model, Poisson regression, hierarchical modelling, as well as the ideas of power and reliability in statistics. These ideas are frequently used in data analysis, but within the scope of this book are not individually broad enough to merit their own chapter.

GLM and Poisson Regression

Linear models and their assumptions

Before we talk about the generalized linear model and see how it works, we need to first discuss **linear models**. Linear models, as you probably already know, use regression to model continuous outcome variable Y, using explanatory variable(s) X that can be either continuous or discrete. In order for the Y variable to be properly predicted by the X variable in a linear regression, we have to satisfy four assumptions (Berry, 1993):

1) Linearity — the regression model used to predict the Y values with the X values must form a straight line.

2) Independence — all individuals in the dataset are sampled independently (they all have the same chance of being

selected as the others).

3) Normality — the Y variable values need to have a normal distribution.

4) Heteroskedasticity is absent — the deviation of the Y values from the model must be equal for all the Y values.

Generalized Linear Models (GLM)

Generalized linear models (GLM) offer us a degree of flexibility. They are used to model data in a similar way to linear regression when at least one of three assumptions are violated: linearity, normality and/or heteroskedasticity. There are three components of a GLM (Nelder and Wedderburn, 1972):

1) the linear predictor — GLM contains an expression that is used to predict the output, as in a linear regression. It is in the form of

$\beta\_0 + \beta X\_1 + \ldots$

2) the probability distribution — the distribution of Y, which does not take on a normal distribution.

3) the link function — a function linking the linear predictor with the non-normal probability distribution so that a model can be generated.

Poisson regression

Two of the most commonly used examples of GLM are logistic regression and Poisson regression. Many of you may already know that logistic regression is used to model binomial and categorical data. Poisson regression is a statistical method that's used when the variable of interest (outcome variable) is count data or rate data (count per period of time). Examples of Poisson distribution include the number of new students enrolled in a school each year, the number of people dying from influenza each year, or the number of bankruptcies filed per month. These data are always positive whole numbers.

Poisson regression works similarly to other regressions in the sense that it is used to find statistically significant variables that can influence the output count (e.g. the number of new students enrolling in a school each year may be influenced by an increase in the number of new residents in the catchment area). At this point, some of you may ask me, "April, if count data are just numbers, why can't they be used in a regression?" The answer to this question is that we cannot find a probability distribution to which all count datasets belong.

Let's go through an example of a Poisson regression in R. The dataset we will use is the 'gala' dataset from the *faraway* R package (Johnson and Raven, 1973, as adapted by Julian Faraway). The dataset contains information about the number of different species of plants in the Galapagos Islands.

```
Galapagos = read.csv("Galapagos.csv")
plantspecies <-
    glm(Species ~ Area + Nearest + Adjacent,
    data=Galapagos, family = "poisson")

summary(plantspecies)
```

First we open the file Galapagos.csv. The regression expression is very straightforward using the glm() function, which is automatically available in the R environment. The first argument of this function contains the model Species ~ Area + Nearest + Adjacent where Species (the outcome variable) represents the number of species of plants on each of the Galapagos Islands; Area, Nearest, and Adjacent are the area of each island, distance to the nearest island, and the area of the adjacent island respectively. The data is set as the Galapagos dataset, and 'family' in this case is specified as having a Poisson distribution; the 'family' argument contains the link function which is characteristic of GLM.

We name the model plantspecies. The summary() function applied to the object gives the model's information:

```
call:
glm(formula = Species ~ Area + Nearest + Adjacent, family = "poisson",
    data = Galapagos)

Deviance Residuals:
    Min       1Q    Median       3Q       Max
-10.739   -8.069    -5.004    2.205    25.653

Coefficients:
              Estimate Std. Error z value Pr(>|z|)
(Intercept)  4.180e+00  2.918e-02 143.283  < 2e-16 ***
Area         4.302e-04  1.199e-05  35.879  < 2e-16 ***
Nearest      5.889e-03  1.434e-03   4.106 4.03e-05 ***
Adjacent    -8.064e-05  2.793e-05  -2.887  0.00389 **
---
Signif. codes:  0 '***' 0.001 '**' 0.01 '*' 0.05 '.' 0.1 ' ' 1

(Dispersion parameter for poisson family taken to be 1)

    Null deviance: 3510.7  on 29  degrees of freedom
Residual deviance: 2587.2  on 26  degrees of freedom
AIC: 2756.1

Number of Fisher Scoring iterations: 6
```

If we look at the coefficients section of the outputs, the p values of all three variables are below 0.05, therefore all variables significantly influence the output variable of the number of plant species in the Galapagos Islands. In Poisson regression, the coefficients are interpreted in a similar way as a logistic regression where we need to use the exponent of the coefficients to determine the influence of the independent variables on the outcome variable. For example, for the area of each of the islands, the coefficient is 4.302e-04 (approximately 0.0004302), which means each increase of the island area by 1 will increase the plant species number by a factor of $e^{0.0004302} = 1.0004303$. * The outputs also show the AIC value of this model, which is used to determine how well models fit the dataset; if we had more than one poisson model of the same outcome but with different independent variables, the model with the lower AIC would be the model with the better fit.

The family entry of the glm() function controls the link function of the GLM. It can have multiple options for different types of GLM. For logistic regression, the family entry is "binomial" and for a linear regression, the family entry is "gaussian" (for normal distribution). The Galapagos dataset has another variable, Endemics, for the number of endemic species on the islands. This can be used in a model along with the other independent variables in the dataset and is a good place to start practicing using Poisson regression. To try out different link functions, use options of the family argument with suitable datasets.

---

*To get rid of scientific notion, code `options(scipen = 999)`. To return to default, use `options(scipen = 0)` (zero).

Hierarchical Modeling

GLMs are used to cope with the violation of the assumptions of linearity, normality, and/or heteroskedasticity. Hierarchical modeling is a method that is used when the *independence* assumption is violated in the data (Gelman and Hill, 2006), put another way, dealing with cluster data. This method is especially handy when it is not feasible to recruit individuals in a dataset independently.

For example, if we want to look at the math test scores in a group of students in a school district, we can sample a few schools, then in the schools we sample a few classes and within the few classes we sample the students. If we collect samples like this, we can see that the dataset is nested within itself (through different schools and classes, because students from the same class or school will be likely to be more similar to each other than from different classes and schools). This gives the dataset a hierarchical feeling. The hierarchical modeling method is used to model data like this while taking into account the student similarities in the dataset by hierarchies such as classes and schools; because of this, they are often referred to as **nested models**, **multilevel models**, or **mixed effect models**.

Some of you might say to yourself, "isn't mixed effect model the method that was used in the longitudinal analysis chapter?" You are right! The mixed effects method that is used to analyze longitudinal data is a variant of hierarchical modelling, where the data at each timepoint are nested within each individual in the dataset. In this portion of the book, let's look at hierarchical modeling without the repeated measures!

The file Education.csv contains an artificially created dataset inspired by the results of a study by Romel *et al* (2020). This indicated that the academic performance of university students as measured by the CPGA was influenced in particular by the 'Big Five' personality trait Openness, and to a lesser extent those of Agreeability and Conscientiousness. The sample contains 1200 students, attending 6 universities, with 4 classes within each university. In this case, the classes are nested in the universities, and the data of the academic and psychological attributes of the students (CPGA, openness, agreeability, conscientiousness) are nested in the classes of each university. The analysis method for this dataset is almost the same as the longitudinal analysis method, through linear mixed modelling.

```
install.packages("lme4"); library(lme4)
Education = read.csv("Education.csv")
head(Education)
```

```
  X id   CGPA  open agree conscience class university
1 1   1 95.594 81.77 85.21      85.42     a    Harris
2 2   2 71.316 80.09 68.87      75.35     b   Johnson
3 3   3 98.890 80.95 74.50      72.09     a   Johnson
4 4   4 47.504 48.15 54.21      59.39     c    Harris
5 5   5 68.775 64.94 70.75      32.17     a    Harris
6 6   6 43.312 41.30 54.14      79.91     a   Allenby
```

First we install and open the lme4 package into R, then we read the dataset into R using read.csv() and name the dataset Education. Each individual (id) has a score for their different attributes and belongs to a class, itself belonging to a university. Note that the categorical variables 'class' and 'university' take letters, not numbers; this is necessary for them to show up in later output.

We then figure out what are the fixed and random effects. Let us say that we know that openness, agreeability, and conscientiousness are population level variables which can influence student CGPA scores, so we set these as fixed variables. We also want to see if students from different classes within different universities can affect CGPA scores, but since the students' level of openness, agreeability and conscientiousness could differ between students from different universities and the different classes of each university, class and university are random variables.

```
Education.mod1 <-
    lmer(CGPA ~ open + agree + conscience
    + (1|class) + (1|university),
    data = Education)
```

Let us build the first of two models. Just as in longitudinal analysis, we use the lmer() function from the lme4 package. First we enter the outcome variable CGPA and the fixed effects 'open', 'agree', and 'conscience', then we enter the random effects. For the first model, we can enter the random intercepts for class and university in the form of (1|class) and (1|university). In this model, named Education.mod1, we want to see if membership of different classes and of different universities *in general* makes a difference to students' CGPA scores.

```
summary(Education.mod1)
```

```
Linear mixed model fit by REML ['lmerMod']
Formula: CGPA ~ open + agree + conscience + (1 | class) + (1 | university)
   Data: Education

REML criterion at convergence: 10596.3

Scaled residuals:
     Min       1Q   Median       3Q      Max
-2.09779 -0.83000  0.03728  0.85885  2.07532

Random effects:
 Groups     Name          Variance Std.Dev.
 university (Intercept)     0.433    0.658
 class      (Intercept)     1.682    1.297
 Residual                 395.597   19.890
Number of obs: 1200, groups:  university, 6; class, 4

Fixed effects:
            Estimate Std. Error t value
(Intercept) 46.68242    2.96232  15.759
open         0.18778    0.03004   6.251
agree        0.06702    0.02976   2.252
conscience   0.04436    0.02968   1.495

Correlation of Fixed Effects:
           (Intr) open   agree
open       -0.452
agree      -0.449 -0.204
conscience -0.481 -0.124 -0.118
```

The intercept under fixed effects is 46.68242. This is the average CGPA score of the entire population of students under this model. Under the random intercepts of university and class, the standard deviations are 0.658 and 1.297. This means that there is a difference in CGPA of 0.658 and 1.297 compared to the average CGPA for students from different universities and classes respectively. For the fixed effects, the results show that students with openness higher by 1 had 0.18778 increase in CGPA while students' agreeability and conscientiousness scores higher by 1 had 0.06702 and

0.04436 increase in CGPA respectively (note that a negative figure would mean a decrease). *

```
Education.mod2 <-
    lmer(CGPA ~ open + agree + conscience
    + (0 + class|university),
    data =Education)
```

Whereas the previous model looks at how class and university as random variables separately influence the CGPA scores of students, we now look at how different classes *within* universities influence the students' CGPA scores. This is represented by the expression (0+class|university). (The warning message about fit singularity after the creation of the model indicates an overfitted model; however, we will overlook this for the purpose of the example and continue to the summary() function.)

```
summary(Education.mod2)
```

---

* To shorten the output figures, code `options(digits = 3)` or whatever is desired. To return to default, use `options(digits = 7)`.

```
Linear mixed model fit by REML ['lmerMod']
Formula: CGPA ~ open + agree + conscience + (0 + class | university)
   Data: Education

REML criterion at convergence: 10596.1

Scaled residuals:
    Min      1Q   Median      3Q     Max
-2.02587 -0.82452  0.04757  0.85527  2.10682

Random effects:
 Groups     Name   Variance  Std.Dev. Corr
 university classa   0.00000  0.0000
            classb   7.10394  2.6653   NaN
            classc   0.09079  0.3013   NaN 1.00
            classd   3.36302  1.8339   NaN 1.00 1.00
 Residual          395.10464 19.8772
Number of obs: 1200, groups:  university, 6

Fixed effects:
            Estimate Std. Error t value
(Intercept) 47.28435    2.89845  16.314
open         0.18734    0.03002   6.240
agree        0.06706    0.02971   2.257
conscience   0.04498    0.02965   1.517

Correlation of Fixed Effects:
           (Intr) open   agree
open       -0.461
agree      -0.457 -0.204
conscience -0.492 -0.124 -0.117
optimizer (nloptwrap) convergence code: 0 (OK)
boundary (singular) fit: see help('isSingular')
```

From the output of this new model, we can see that the intercept under 'Fixed effects' is 47.284. If we look at the random effects of classa, classb, classc, and classd, we can see from the standard deviation that there is a difference in CGPA of 0.0000, 2.665, 0.301, and 1.834 from the average CGPA score. For the fixed effects, the results show that students' traits of openness, agreeability and conscientiousness, as each was higher by 1, had a respective increase in CGPA of 0.18734, 0.06706 and 0.04498 (again, negative scores would indicate a decrease in CGPA score).

Power Analysis

Statistics cannot exist without the principles of power and reliability. Power, from the statistical perspective, means the likelihood of making a mistake when we use a statistical result to decide whether or not to reject a null hypothesis. When testing a null hypothesis, we may make Type I and Type II errors. Type I, sometimes represented as α (alpha), is the likelihood of calculating a statistic value that points towards rejecting the null hypothesis when in truth we should accept it. Type II, sometimes represented as β (beta), is the likelihood of calculating a statistical value that points towards not rejecting the null hypothesis when in truth it should be rejected.

In studies, there are usually more consequences to wrongly accepting the null hypothesis. Therefore the power of a study (which is also the probability of falsely rejecting the null hypothesis, the direct opposite of not rejecting the null hypothesis when it's true) is measured as $1-\beta$, the likelihood of not making a Type II error (Banerjee *et al*, 2009). This calculation is used when we say a study has 95% power, meaning that there is a 95% chance that a Type II error will not happen.

In statistics, we like to use the idea of power when we are trying to calculate the proper sample size to ensure the trustworthiness of study results. When we are using the critical value of $p < 0.05$ as a standard to accept or reject null hypothesis, what we are really saying is that "if the dataset and its results from the analysis has less than a 5% chance of occurring under the null hypothesis, then we reject the null hypothesis. If it is more than 5%, we should (for now) accept the null hypothesis".

To make things a little clearer, let's say that we have a study with a significance level of 0.05. It does not automatically mean that the power of the study is 95%. The significance level is the probability of Type I error; it is separate from the power of the study.

```
install.packages("pwr")
library(pwr)
```

To understand how to use power to calculate sample size, let's go through an example in R. A great package for calculating sample size for simple statistical methods is the *pwr* package. The functions for different statistical tests include:

pwr.2p.test()   — two proportions (same sample sizes)
pwr.2p2n.test()   — two proportions (different sample sizes)
pwr.anova.test()   — balanced (equal number of observations) one-way ANOVA
pwr.chisq.test()   — chi-squared tests
pwr.f2.test()   — general linear model
pwr.p.test()   — proportion tests (one sample)
pwr.r.test()   — correlation test
pwr.t.test()   — *t* tests (one sample, two samples, and paired)
pwr.t2n.test()   — *t* tests (two samples, different sizes)

```
?  pwr.t.test
```

If we type in any of these functions using their names preceded by a question mark, we get a description of what the function does, and its arguments (this works for all built-in functions in R and all R packages).

```
pwr.t.test(n = NULL, d = 0.5,
    sig.level = 0.05, power = 0.95)
```

Let's say that we want to calculate the sample size for a study where we have to compare the means of two samples using a t test; we can use the pwr.t.test() function. Since the sample size is what we are looking for, we put in n = NULL. Here, we have set the significance level and power at 0.05 and 0.95 respectively (these could be changed depending on the requirements of your study). The 'd' argument refers to Cohen's d, which is a value used to test the magnitude of the differences between the means of the two samples (0.1 is considered small, 0.3 medium, 0.5 large); for our example, let's use 0.5. If we run our function, for sample size n, we need 105 (rounded up) individuals per sample:

```
     Two-sample t test power calculation

              n = 104.9279
              d = 0.5
      sig.level = 0.05
          power = 0.95
    alternative = two.sided

NOTE: n is number in *each* group
```

For more advanced level statistical methods, statisticians often rely on the 'law of large numbers' and discipline-specific knowledge to determine the sample size. But if your current study uses simple statistical methods like the *t* test, chi-square test, or one-way ANOVA, the pwr package can be very useful. Why not try out other functions in the package to calculate the sample size for a study of your own design?

Reliability

Like power, reliability is also an important concept in testing for the quality of a dataset. It is used to measure the consistency of the data. If data is not consistent, attributes may not be accurately measured, and then even the most powerful of tools may mislead the researcher.

Here are three types of reliability which may be tested statistically:

1) Test-retest reliability — consistency of measurement in the same variables, over time.

2) Internal reliability — consistency of results for different variables.

3) Inter-rater reliability — consistent measurements (ratings) are provided by different people.

Test-retest reliability

Test-retest reliability requires us to look at the consistency of the same measurements over time (Lowe and Rabbitt, 1998). For an example where we believe that a result will stay very similar, let us measure the body temperatures of 10 people on two consecutive days, at the same time of day; these are very unlikely to change significantly. To measure test-retest reliability, we can test for the level of correlation between measurements of the same people on two different occasions using Pearson's test, showing the correlation statistic r.

```
BodyTemp1 <- c(96.24571, 97.15269,
    96.34844, 96.59914, 96.97056, 97.46070,
    95.89002, 96.35390, 97.13040, 95.65425)
BodyTemp2 <- c(96.32290, 96.39820,
    95.90986, 96.49517, 96.91765, 97.47595,
    96.79387, 97.06743, 97.48383, 97.39397)
```

So we have two vectors, BodyTemp1 and BodyTemp2, representing the first and second days, each containing data for the same 10 people.

```
BodyTemp <- data.frame(BodyTemp1, BodyTemp2)
BodyTemp
```

We can put both of these vectors into a dataframe called BodyTemp, consisting of two columns:

```
   BodyTemp1 BodyTemp2
1   96.24571  96.32290
2   97.15269  96.39820
3   96.34844  95.90986
4   96.59914  96.49517
5   96.97056  96.91765
6   97.46070  97.47595
7   95.89002  96.79387
8   96.35390  97.06743
9   97.13040  97.48383
10  95.65425  97.39397
```

```
PearsonR <- cor(BodyTemp1, BodyTemp2,
   method = "pearson")

PearsonR

[1] 0.1891994
```

To calculate the correlation value, Pearson's r, all we have to do is to place the two columns into the cor() function's first two arguments and specify the method as "pearson" (with data that is not continuous, use a non-parametric test, "spearman"). If we run the new object, PearsonR, by itself, it gives the value 0.19 (rounded). Pearson's r correlation has a value between −1 and 1. For the test-retest reliability to be good, we need a Pearson r correlation of +0.8 or higher. In our example, the test-retest reliability is not good; I hope they get better soon!

Internal reliability

Internal reliability is when we collect data which contains variables that are related to each other in a sample; the contents of the related variables must be consistent with each other. For example, if we have a group of people answering multiple choice questions about themselves about how extraverted they are, the same person should have consistent answers for all the questions. Let's use the dataset Extraversion.csv. It contains the responses of 30 people who were each asked how much they agreed with the following five statements (on a 5-point scale):

- E1 I am the life of the party
- E3 I feel comfortable around people
- E5 I start conversations
- E7 I talk to a lot of different people at parties
- E9 I don't mind being the center of attention

```
install.packages("psych"); library(psych)
Extraversion <- read.csv("Extraversion.csv")
alpha(Extraversion)
```

```
Reliability analysis
Call: alpha(x = Extraversion)

  raw_alpha std.alpha G6(smc) average_r S/N   ase mean  sd median_r
   0.89      0.89       0.9     0.62   8.3 0.032  2.9 1.1    0.59

      95% confidence boundaries
           lower alpha upper
Feldt       0.81  0.89  0.94
Duhachek    0.83  0.89  0.95

 Reliability if an item is dropped:
      raw_alpha std.alpha G6(smc) average_r S/N alpha se  var.r med.r
d.E1    0.85      0.85     0.86     0.59   5.7  0.046 0.0216  0.57
d.E3    0.89      0.89     0.88     0.67   8.3  0.034 0.0154  0.67
d.E5    0.85      0.85     0.84     0.58   5.6  0.047 0.0157  0.59
d.E7    0.88      0.88     0.88     0.64   7.2  0.037 0.0282  0.66
d.E9    0.87      0.87     0.86     0.63   6.8  0.039 0.0079  0.59

 Item statistics
       n raw.r std.r r.cor r.drop mean  sd
d.E1  30  0.88  0.89  0.86  0.82   2.6 1.3
d.E3  30  0.76  0.76  0.69  0.63   3.2 1.3
d.E5  30  0.90  0.89  0.88  0.83   3.0 1.4
d.E7  30  0.81  0.81  0.74  0.69   2.6 1.4
d.E9  30  0.83  0.83  0.80  0.72   3.0 1.4

Non missing response frequency for each item
        1    2    3    4    5 miss
d.E1 0.23 0.27 0.30 0.10 0.10   0
d.E3 0.07 0.30 0.20 0.23 0.20   0
d.E5 0.13 0.27 0.27 0.10 0.23   0
d.E7 0.33 0.17 0.23 0.13 0.13   0
d.E9 0.23 0.10 0.27 0.23 0.17   0
```

To calculate internal consistency, we need to calculate Cronbach's α, otherwise known as *alpha* (Cronbach, 1951), using the alpha() function from the *psych* package. We just read the Extraversion dataset into R using read.csv() and place the dataset within the alpha() function.

In the output, the **std.alpha** value is 0.89. A Cronbach's α of +0.8 or higher shows that a dataset has great internal consistency. One important point is that this method will only work when the variable values are positively correlated. For this dataset, if we had thrown in a few statements such as "I like to be alone", "I don't like to socialize", or "I am introverted" along with the existing five, Cronbach's α will not work unless we recode the scores of the additional statements so the data all heads in the same direction.

Inter-rater reliability

Inter-rater reliability involves looking at the consistency of measurements for different people rating the same variables in a study. If the data is numerical, we can use Cronbach's α to perform the test. If the data is categorical, we can use a method called Cohen's Kappa (κ) to perform the test (McHugh, 2012).

The R package *irr* contains two particularly useful example datasets:

```
install.packages("irr"); library(irr)
data("anxiety") # contains numerical data
data("diagnoses") # contains categorical data
```

To use these datasets, we must first install the *irr* package and open it in R. The data() function allows direct use of the datasets by their names.

Numerical ratings

```
head(anxiety, n=3) # (truncated)
```

```
   rater1 rater2 rater3
1     3      3      2
2     3      6      1
3     3      4      4
```

The 'anxiety' dataset — numerical — contains the level of anxiety (on a scale of 1-6) for 20 subjects by 3 raters.

```
# library(psych) if not already loaded
alpha(anxiety) # data frame within function
```

The data frame is placed within the alpha() function.

```
Reliability analysis
Call: alpha(x = anxiety)

  raw_alpha std.alpha G6(smc) average_r  S/N  ase mean   sd median_r
     0.45      0.46    0.38      0.22 0.85 0.21  2.9 0.94    0.28

     95% confidence boundaries
            lower alpha upper
Feldt      -0.15  0.45  0.77
Duhachek    0.03  0.45  0.87

 Reliability if an item is dropped:
        raw_alpha std.alpha G6(smc) average_r  S/N alpha se var.r med.r
rater1      0.44      0.44   0.282     0.282 0.78     0.25    NA 0.282
rater2      0.15      0.15   0.083     0.083 0.18     0.38    NA 0.083
rater3      0.46      0.46   0.300     0.300 0.86     0.24    NA 0.300

 Item statistics
        n raw.r std.r r.cor r.drop mean  sd
rater1 20  0.69  0.66  0.37   0.24  3.1 1.5
rater2 20  0.74  0.76  0.57   0.40  3.1 1.3
rater3 20  0.65  0.66  0.35   0.22  2.3 1.3

Non missing response frequency for each item
          1    2    3   4   5    6 miss
rater1 0.10 0.30 0.25 0.1 0.2 0.05    0
rater2 0.05 0.25 0.40 0.2 0.0 0.10    0
rater3 0.35 0.25 0.25 0.1 0.0 0.05    0
```

The std.alpha value is 0.46, way lower than 0.8. So the anxiety dataset does not have great interrater consistency.

Categorical ratings

```
View(diagnoses) # truncated image
```

	rater5	rater6
;	4. Neurosis	4. Neurosis
	5. Other	5. Other
renia	3. Schizophrenia	5. Other
	5. Other	5. Other
;	4. Neurosis	4. Neurosis
renia	3. Schizophrenia	3. Schizophrenia
reni	5. Other	5. Other

The dataset — categorical — contains psychiatric diagnoses (Depression, Personality Disorder, Schizophrenia, Neurosis, Other) of 30 subjects by 6 raters.

```
library(irr)
kappa2(diagnoses[, c("rater5", "rater6")],
    weight = "unweighted")
```

For this dataset, we use the kappa2() function from the *irr* package to calculate Cohen's Kappa. The function can only compare the interrater consistency of 2 raters at once and for our example, we choose the ratings of rater 5 and rater 6. We can place the rater5 and rater6 columns (specified using a vector within the dataframe) in the first argument of the kappa2 function and specify weight as "unweighted".

```
Cohen's Kappa for 2 Raters (Weights: unweighted)

Subjects = 30
  Raters = 2
   Kappa = 0.648

       z = 5.38
 p-value = 7.36e-08
```

If we look at the output, the kappa value is 0.65 (rounded). For Cohen's Kappa and how it indicates agreement between the ratings of two raters:

≤ 0 — no agreement
0.01–0.20 — none to slight
0.21–0.40 — fair
0.41– 0.60 — moderate
0.61–0.80 — substantial
0.81–1.00 — almost perfect agreement

So a kappa value of 0.65 indicates substantial agreement between the ratings of rater 5 and rater 6, meaning good interrater consistency for those two raters in this dataset.

Further reading

For generalized linear models, Poisson regression, and hierarchical modeling, I would suggest Gelman and Hill (2006). Useful coverage of power analysis may be found in Pedahazur and Schmelkin (1991), and in Stevens (2002). Some worthwhile reading on reliability may be found in Pedahazur and Schmelkin (1991), and Strube (2000).

References

Ballinger G A (2004)   Using Generalized Estimating Equations for Longitudinal Data Analysis. *Organizational Research Methods*, *7*(2), 127–150.
< https://doi.org/10.1177/1094428104263672 >

Banerjee A, Chitnis U, Jadhav S, Bhawalkar J, and Chaudhury S (2009) Hypothesis testing, type I and type II errors. *Industrial Psychiatry Journal*, *18*(2), 127.
< https://doi.org/10.4103/0972-6748.62274 >

Berry W D (1993) *Understanding Regression Assumptions* (1st ed.). SAGE.

Bland J M and Altman D G (2004) The logrank test. *BMJ*, *328*(7447), 1073.
< https://doi.org/10.1136/bmj.328.7447.1073 >

Boashash B (2015) *Time-Frequency Signal Analysis and Processing: A Comprehensive Reference* (2nd ed). Cambridge, Mass: Academic Press.

Bollen K A and Curran P J (2006) *Latent Curve Models: A Structural Equation Perspective*. Hoboken, NJ: Wiley.

Borg I and Groenen P (2005) *Modern Multidimensional Scaling: Theory and Applications* (2nd ed.). Springer.

Box G E P, Jenkins G M and Reinsel G C (1994) *Time Series Analysis: Forecasting and Control* (3rd ed). Englewood Cliffs, NJ: Prentice Hall.

Bryant F B and Yarnold P R (1995) 'Principal-Components Analysis and Exploratory and Confirmatory Factor Analysis', in Grimm and Yarnold (1995).

Collett D (2003) *Modelling Survival Data in Medical Research* (2nd ed.). London: Chapman & Hall/CRC.

Costa P and McCrae R R (1992) 'Normal personality assessment in clinical practice: The NEO Personality Inventory.' *Psychological Assessment*, *4*, 5–13.

Cox D and Oakes D (1984) *Analysis of Survival Data.* London: Chapman & Hall.

Croissant Y (2022) R package *plm* (2.6-1)

Cronbach L J (1951) Coefficient alpha and the internal structure of tests. *Psychometrika, 16*(3), 297-334. <https://doi.org/10.1007/BF02310555>

Davis C (2019) *Statistical testing with jamovi and JASP open source software - Psychology*. Norwich: Vor Press.

Davis C (2022) *Statistical Testing With R* (second edition). Norwich: Vor Press.

Diggle P J, Heagerty P, Liang K-Y, and Zeger S L (2002) *Analysis of longitudinal data* (2nd ed.). Oxford University.

Everitt B and Hothorn T (2011) *An Introduction to Applied Multi-variate Analysis with R* (1st ed.). New York: Springer.

Fitzmaurice G M, Laird N M, and Ware J H (2004) *Applied Longitudinal Analysis*. John Wiley & Sons.

Fox J and Weisberg S (2019) *Cox Proportional-Hazards Regression for Survival Data in R*. Thousand Oaks CA: Sage.

Friedman J H (1989) Regularized Discriminant Analysis. *Journal of the American Statistical Association, 84*, 165-75.

Gelman A and Hill J (2006) *Data Analysis Using Regression and Multilevel/Hierarchical Models* (1st ed.). Cambridge University Press.

Girden E (1992) *ANOVA: Repeated measures* (1st ed.). Sage.

Greene W (2002) *Econometric Analysis* (5th ed.) New Jersey: Prentice-Hall.

Greenhouse S W and Geisser S (1959) On methods in the analysis of profile data. *Psychometrika, 24*, 95–112.

Greenwood M (1926) A report on the natural duration of cancer. *Reports on Public Health and Medical Subjects, 33*

Grimm L G and Yarnold P R (1995) *Reading and Understanding Multivariate Statistics*. Washington D.C.: American Psychological Association.

Grimm L G and Yarnold P R (2000) *Reading and Understanding More Multivariate Statistics*. Washington D.C.: American Psychological Association.

Helliwell J, Layard R, and Sachs J, eds (2015) World Happiness Report 2015. New York: Sustainable Development Solutions Network.

Helliwell J, Layard R, and Sachs J, eds (2019) World Happiness Report 2019. New York: Sustainable Development Solutions Network.

Hosmer D and Lemeshow S (1999) *Applied Survival Analysis: Regression Modeling of Time to Event Data*. New York: John Wiley and Sons.

Hotelling H (1936) Relations Between Two Sets of Variates. *Biometrika, 28*, 321–377.
< https://doi.org/10.1093/biomet/28.3-4.321 >

Hyndman R and Athanasopoulos G (2018) *Forecasting: principles and practice* (2nd ed). Melbourne: OTexts.

James G, Witten D, Hastie T, and Tibshirani R (2021) *An Introduction to Statistical Learning: with Applications in R* (2nd ed). Springer.

Johnson M P and Raven P H (1973) Species Number and Endemism: The Galápagos Archipelago Revisited. *Science, 179*(4076), 893–895.
< https://doi.org/10.1126/science.179.4076.893 >

Kaplan E L and Meier P (1958) Nonparametric Estimation from Incomplete Observations. *Journal of the American Statistical Association*, *53*(282), 457.
< https://doi.org/10.2307/2281868 >

Klecka W (1980) *Discriminant Analysis*. SAGE.

Kleinbaum D G and Klein M (2006) *Survival Analysis: A Self-Learning Text (Statistics for Biology and Health)* (2nd ed.) New York: Springer.

Klem L (1995) 'Path Analysis', in Grimm and Yarnold (1995).

Klem L (2000) 'Structural Equation Modeling', in Grimm and Yarnold (2000)

Kwiatkowski D, Phillips P, Schmidt P, and Shin Y (1992) 'Testing the Null Hypothesis of Stationarity Against the Alternative of a Unit Root', *Journal of Econometrics*, *1*(3), 159–178.

Laird N M and Ware, J. H. (1982) Random-effects models for longitudinal data. *Biometrics*, *38*(4), 963–974.

Liang K-Y and Zeger S L (1986) Longitudinal data analysis using generalized linear models. *Biometrika*, *73*(1), 13–22.
< https://doi.org/10.1093/biomet/73.1.13 >

Lishinski A (2018) 'lavaanPlot: Path Diagrams for Lavaan Models via DiagrammeR', R package version 0.5.1),
< https://cran.r-project.org/package=lavaanPlot >

Lowe C and Rabbitt P (1998) Test/re-test reliability of the CANTAB and ISPOCD neuropsychological batteries: theoretical and practical issues.*Neuropsychologia,36*(9),915–923.

< https://doi.org/10.1016/S0028-3932(98)00036-0 >

Luke S G (2017) Evaluating significance in linear mixed-effects models in R. *Behavior Research Methods, 49*(4), 1494–1502.
< https://doi.org/10.3758/s13428-016-0809-y >

McHugh M L (2012) Interrater reliability: the kappa statistic. *Biochemia Medica, 22*(3), 276–282.

Nelder J A and Wedderburn R W M (1972) Generalized Linear Models. *Journal of the Royal Statistical Society. Series A (General), 135*(3), 370. < https://doi.org/10.2307/2344614 >

OSMI (2016) Mental Health in Tech Survey 2016.
<https://www.kaggle.com/osmi/mental-health-in-tech-2016>

Pedahazur E J and Schmelkin L P (1991) Measurement, Design, and Analysis. HillsDale, NJ: Lawrence Erlbaum.

Pennsylvania State University (2022) *STAT 505 Applied Multivariate Statistical Analysis, 13.2* - Example: Sales Data.
< https://online.stat.psu.edu/stat505/lesson/13/13.2 >

Richards S J (2012) A handbook of parametric survival models for actuarial use. *Scandinavian Actuarial Journal, 2012(4)*, 233–257.
< https://doi.org/10.1080/03461238.2010.506688 >

Romel J, Rehana J, and Zia-ur-Rehman R (2020)   The Big Five Personality Traits And Academic Performance. *Journal of Law and Social Studies, 2*, 10-19.

Rosseel Y (2012) 'lavaan: An R Package for Structural Equation Modeling', *Journal of Statistical Software, 48*(2),

1–36. < http://www.jstatsoft.org/v48/i02/ >

Schumacker, R E and Lomax, R G (2010) *A Beginner's Guide to Structural Equation Modeling*, 3rd ed. New York: Routledge.

Shumway R and Stoffer D (2017) *Time Series Analysis and Its Applications: With R Examples* (4th ed). New York: Springer.

Silva A P D and Stam A (1995) 'Discriminant Analysis', in Grimm and Yarnold (1995).

Singh A (2018) *A Gentle Introduction to Handling a Non-Stationary Time Series in Python*. Retrieved March 20, 2021, from
< https://www.analyticsvidhya.com/blog/2018/09/non-stationary-time-series-python/ >

Stalans L J (1995) 'Multidimensional Scaling', in Grimm and Yarnold (1995).

Stevens J P (2002) *Applied Multivariate Statistics for the Social Sciences*. Mahwah, NJ: Lawrence Erlbaum.

Strube M J (2000) Reliability and Generalizability, in Grimm and Yarnold (2000).

Thompson B (2000) 'Canonical Correlation Analysis', in Grimm and Yarnold (2000).

U.C. San Diego (2016) *Carbon Dioxide Levels in the Atmosphere*. Retrieved November 30, 2020, from
< https://www.kaggle.com/ucsandiego/carbon-dioxide >

Index

CFA, viii, 11-32, 45
>   basic CFA, 12-22
>   multiple groups CFA, 22-32

Cohen's Kappa (κ), 221, 223-4

confirmatory factor analysis, see CFA

constant mean, 50, 51

constant variance, 50-51

Cox model (Cox proportional hazard model), ix, 99, 119-30

Cronbach's α (alpha), 220-22

Dickey-Fuller test, see ADF

discriminant analysis, x, 165-79
>   assumptions, 167-9;
>   linear discriminant analysis (LDA), 170-74
>   quadratic discriminant analysis (QDA), 174-6
>   regularized discriminant analysis (RDA), 176-179

'drift', 76

decomposition, 55, 68-9

deviance residuals, 120-21, 127-9

dfbeta values, see deviance residuals

differencing, 55, 67-8

events (in survival analysis), 96

exponential distribution, 132, 136-8

fixed effects, 153, 154, 210, 213

GEE (generalized estimating equations), ix, 159-63

generalized estimating equations, see GEE

generalized linear models, see GLM

GLM (generalized linear models), x, 203-4

MDS, x, 189-200 (metric, 189-95; non-metric, 195-200)
mixed model, see linear mixed effects model,
     also hierarchical modeling
moving average, 48-49
multidimensional scaling, see MDS
multilevel modeling, see hierarchical modeling,
     also linear mixed effects model
multi-outcome analysis, see multivariate analysis
multivariate analysis, 165, 201, ch. 5 *passim*
nested models, see hierarchical modeling
NNFI (aka Tucker Lewis index), 29
outliers (Cox model), 120-1
PACF (partial autocorrelation function), 70-71, 72,
   83-9, 92, 93
panel analysis, see longitudinal analysis
partial autocorrelation function, see PACF
path analysis, viii, 4-11, 45
Poisson regression, x, 204-7
power analysis, x-xi, 214-17
prediction (in times series analysis), 76-8
product limit method, see Kaplan-Meier
proportional hazards, 119-20, 125
random effects, 153, 154-5, 210
reliability, xi, 217-24
     internal reliability, 219-21
     inter-rater reliability, 221-24
     test-retest reliability, 217-19

repeated measures analysis, see longitudinal analysis
repeated measures ANOVA, 143-4
      one-way, 145-8
      two-way, 148-52
RMSEA, 29
rolling means, 54-5, 59-62, 65-6
SARIMA, 79-93
Schoenfeld residuals, 120, 125-6
scree plot (for MDS), 193-4, 198-9
seasonal ARIMA, see SARIMA
SEM (structural equation modeling)
      umbrella term, viii, 2-3, ch. 1 *passim*
      specific technique (basic), viii, 32-8, 45
Shapiro-Wilk test, 143, 146, 149-50, 167-8
smoothing, see rolling means
stationary vs non-stationary data, 49-69
structural equation modeling, see SEM
survival analysis (in general), ix, 96-100, 138-9, ch. 3 *passim*
time series analysis, viii-ix, 47-49, ch. 2 *passim*
time-to-event analysis, see survival analysis
Tucker Lewis index, see NNFI
Wald test (for Cox model), 123
Weibull distribution, 131-6
within-subjects ANOVA, see repeated measures ANOVA

www.ingramcontent.com/pod-product-compliance
Lightning Source LLC
Chambersburg PA
CBHW071553210326
41597CB00019B/3221